Intermediate 1

Biology

2004 Exam

2005 Exam

2006 Exam

2007 Exam

2008 Exam

Leckie × Leckie

First exam published in 2004.
Published by Leckie & Leckie Ltd, 3rd Floor, 4 Queen Street, Edinburgh EH2 1JE
tel: 0131 220 6831 fax: 0131 225 9987 enquiries@leckieandleckie.co.uk www.leckieandleckie.co.uk

ISBN 978-1-84372-647-0

A CIP Catalogue record for this book is available from the British Library.

Leckie & Leckie is a division of Huveaux plc.

Leckie & Leckie is grateful to the copyright holders, as credited at the back of the book, for permission to use their material. Every effort has been made to trace the copyright holders and to obtain their permission for the use of copyright material. Leckie & Leckie will gladly receive information enabling them to rectify any error or omission in subsequent editions.

2004 | Intermediate I

[BLANK PAGE]

FOR OFFICIAL USE

Section B Total

X007/101

NATIONAL QUALIFICATIONS 2004

WEDNESDAY, 19 MAY
9.00 AM – 10.30 AM

BIOLOGY
INTERMEDIATE 1

Fill in these boxes and read what is printed below.

Full name of centre

Town

Forename(s)

Surname

Date of birth
Day Month Year

Scottish candidate number

Number of seat

SECTION A

Instructions for completion of Section A are given on page two.

SECTION B

1. All questions should be attempted.
2. The questions may be answered in any order but all answers are to be written in the spaces provided in this answer book, and must be written clearly and legibly in ink.
3. Additional space for answers and rough work will be found at the end of the book. If further space is required, supplementary sheets may be obtained from the invigilator and should be inserted inside the front cover of this book.
4. The numbers of questions must be clearly inserted with any answers written in the additional space.
5. Rough work, if any should be necessary, should be written in this book and then scored through when the fair copy has been written.
6. Before leaving the examination room you must give this book to the invigilator. If you do not, you may lose all the marks for this paper.

LIB X007/101 6/3570

Read carefully

1 Check that the answer sheet provided is for Biology Intermediate 1 (Section A).

2 Fill in the details required on the answer sheet.

3 In this section a question is answered by indicating the choice A, B, C or D by a stroke made in **ink** in the appropriate place in the answer sheet—see the sample question below.

4 For each question there is only **one** correct answer.

5 Rough working, if required, should be done only on this question paper—or on the rough working sheet provided—**not** on the answer sheet.

6 At the end of the examination the answer sheet for Section A **must** be placed **inside** this answer book.

Sample Question

Which of the following foods contains a high proportion of fat?

A Bread

B Butter

C Sugar

D Apple

The correct answer is **B**—butter. A **heavy** vertical line should be drawn joining the two dots in the appropriate box in the column headed **B** as shown in the example on the answer sheet.

If, after you have recorded your answer, you decide that you have made an error and wish to make a change, you should cancel the original answer and put a vertical stroke in the box you now consider to be correct. Thus, if you want to change an answer D to an answer B, your answer sheet would look like this:

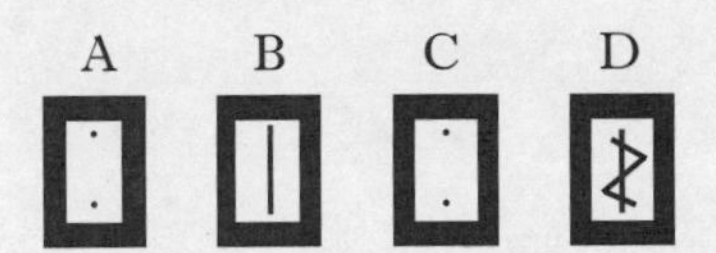

If you want to change back to an answer which has already been scored out, you should enter a tick (✓) to the **right** of the box of your choice, thus:

A B C D **or** A B C D

SECTION A

All questions in this Section should be attempted.

Answers should be given on the separate answer sheet provided.

Questions 1 and 2 refer to the following diagram.

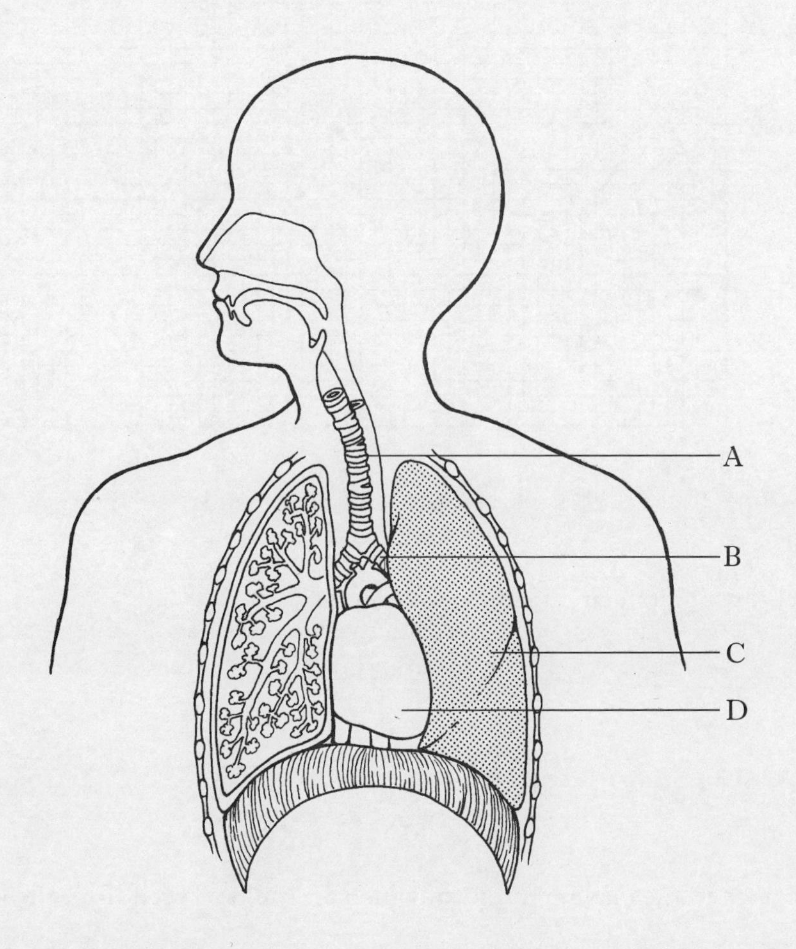

1. Which structure is the windpipe?

2. The function of part D is to

A remove carbon dioxide from the blood

B carry oxygen around the body

C pump blood around the body

D remove oxygen from the air.

[Turn over

3. The graph below shows the volume of air in the lungs of a student during a 30 second period of time.

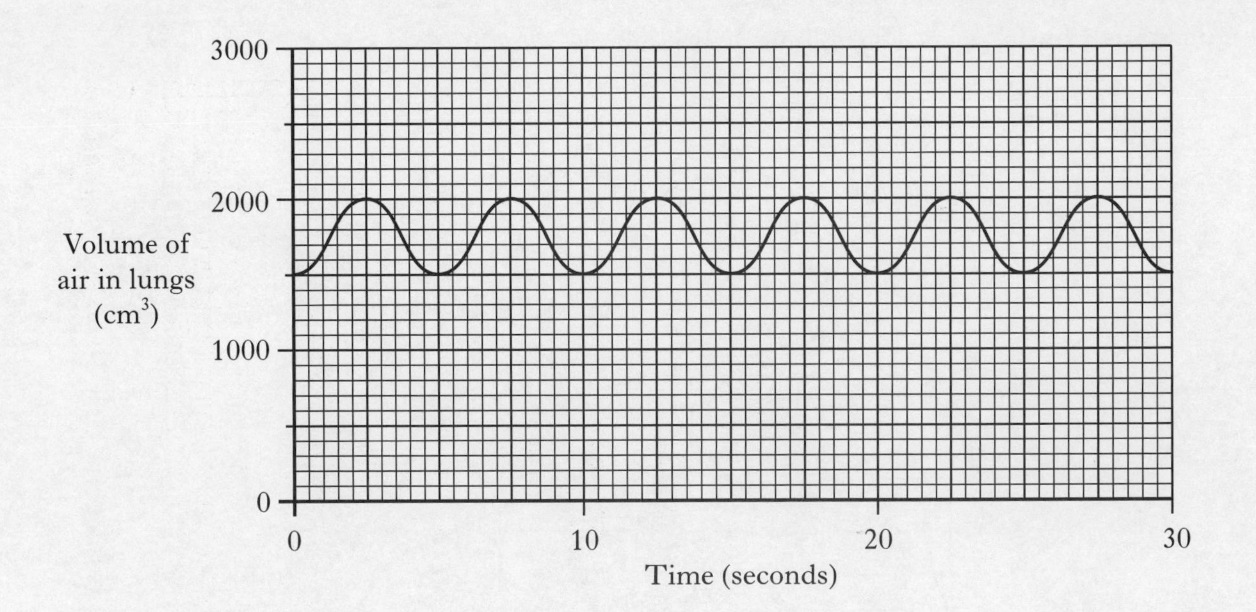

The student's breathing rate, in breaths per minute is

A 6

B 12

C 13

D 24.

4. What name is given to the maximum rate at which air can be forced from the lungs?

A Breathing rate

B Vital capacity

C Tidal volume

D Peak flow

5. The resting pulse rates and recovery times after exercise for four students are shown below.

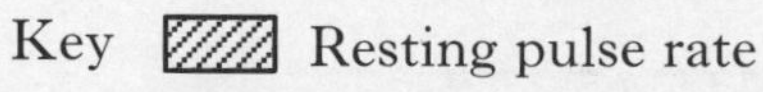
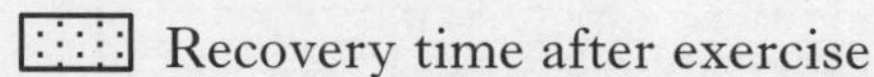
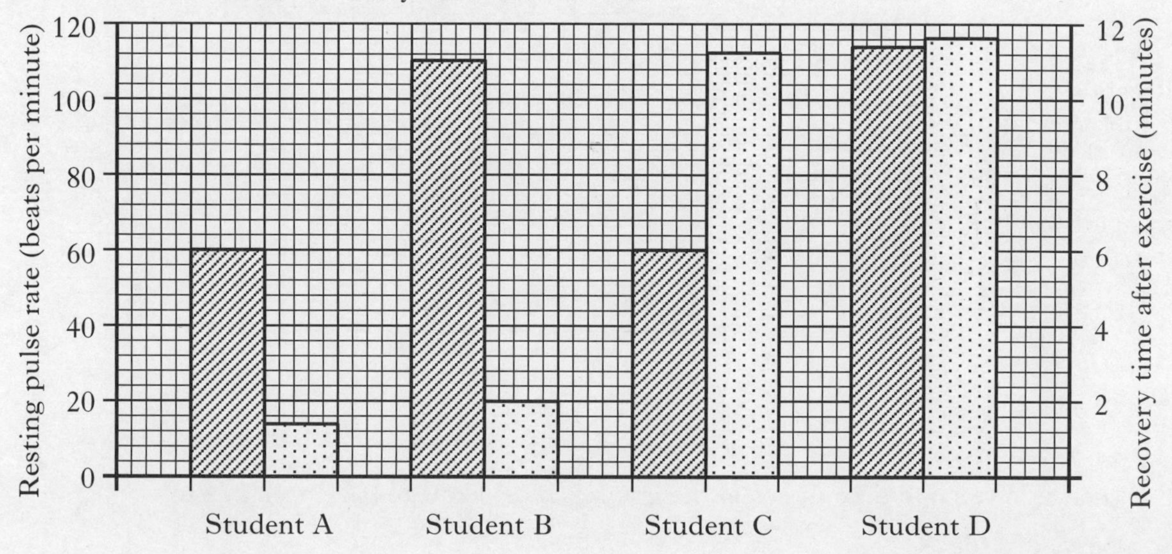

Which student is most likely to be the fittest?

6. What can happen to the size and strength of muscles if they are **not** exercised regularly?

A Size increases and strength decreases.

B Size decreases and strength increases.

C Size and strength both increase.

D Size and strength both decrease.

7. The percentages of different blood groups present in the British population are shown in the table below.

Blood Group	*Percentage of Population*
O	47
A	42
B	8
AB	3

In a British town of 100 000 people, how many would be expected to have blood group A?

A 42

B 2381

C 42 000

D 45 000

8. The graph below shows a student's reaction time over ten trials.

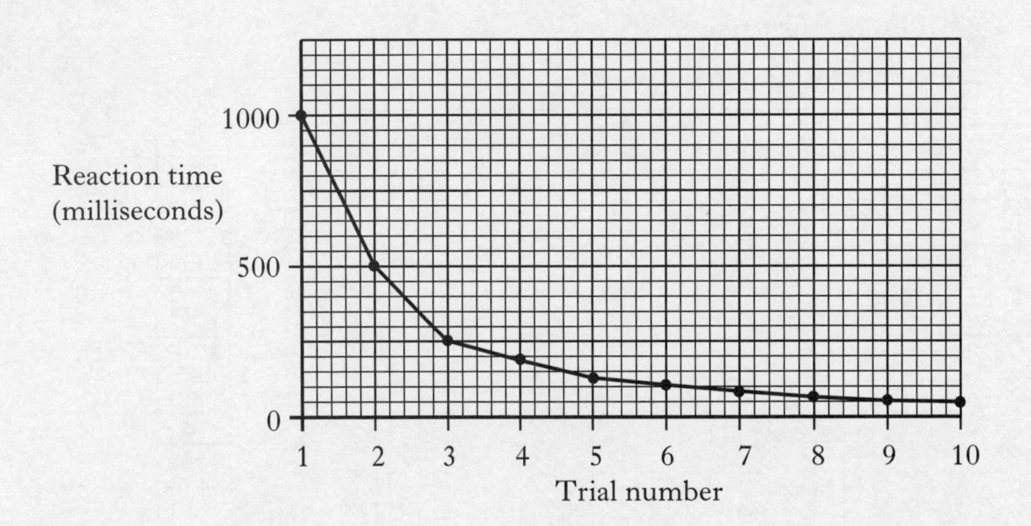

The change in reaction time shown in the graph is most likely to have been due to

A a nerve disorder

B practice

C fatigue

D drinking alcohol.

9. The energy content of different foods is shown in the table below.

Food	*Energy content* (kJ per 100 g)
cornflakes	1500
milk	280
bread	1000
butter	3100

What is the energy content in a snack of 200 g bread, 10 g of butter and 100 g of milk?

A 1590 kJ

B 2590 kJ

C 4380 kJ

D 5380 kJ

10. Which of the following produces the carbon dioxide gas that makes dough rise?

A Bacteria

B Alcohol

C Yeast

D Lactic acid

11. Immobilisation techniques can be used in the production of

A bread

B beer

C cheese

D fermented milk drinks.

12. Which line in the table is true for **brewery** conditioned beer?

	Yeast	*Carbon dioxide*
A	not removed	added
B	removed	not added
C	removed	added
D	not removed	not added

[Turn over

Questions 13 and 14 refer to the following information.

The steps in an investigation into the effects of four antifungal solutions are outlined below.

1. Four agar plates were spread with the same type of fungus.
2. A well was made in the centre of the agar in each plate.
3. A different antifungal solution was placed in each well.
4. The plates were incubated.
5. The size of any clear zone around each well was observed.

The results are shown below.

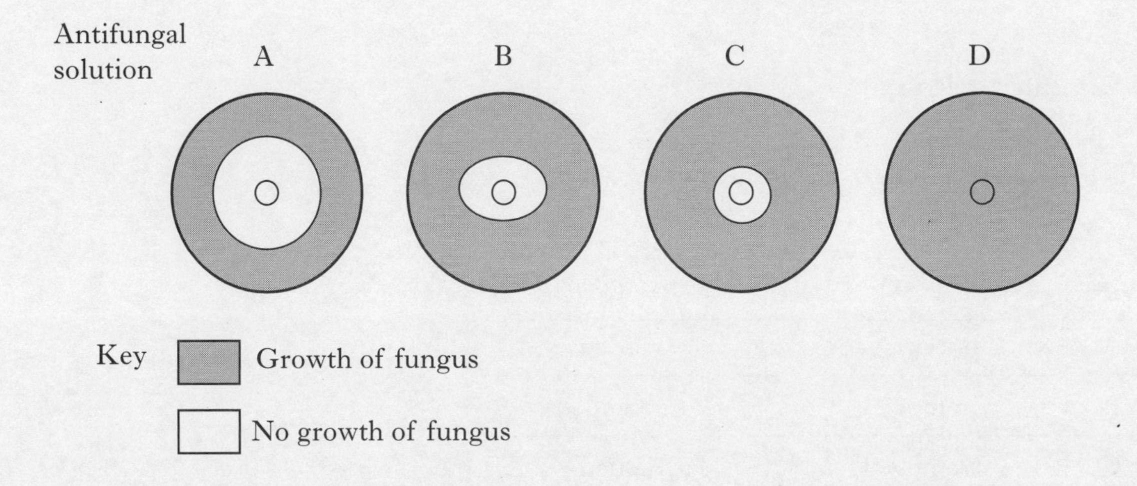

13. Which antifungal solution was **most** effective against the fungus?

14. Which of the following describes how a **control** for this investigation would differ from the plates above?

A Different fungus

B Different antifungal solutions

C Water instead of antifungal solution

D Water instead of fungus

15. Which line in the table describes correctly the effects of disposal of waste whey into rivers?

	Number of bacteria in river	*Oxygen content of river*	*Number and types of other organisms*
A	decreases	increases	increase
B	increases	decreases	increase
C	increases	decreases	decrease
D	decreases	increases	decrease

16. The fermenter below is used in the manufacture of antibiotics.

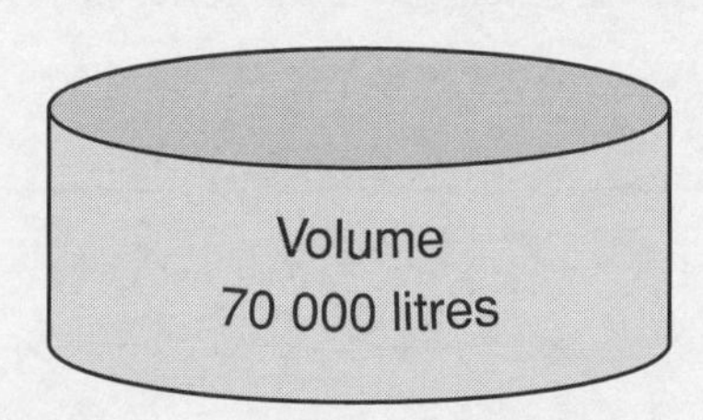

The fermenter produces 5 g of antibiotic **per litre** every day.

The total number of grams of antibiotic produced in **one week** in the fermenter is

A 70 000

B 350 000

C 490 000

D 2 450 000.

Questions 17 and 18 refer to the diagram of a broad bean seed below.

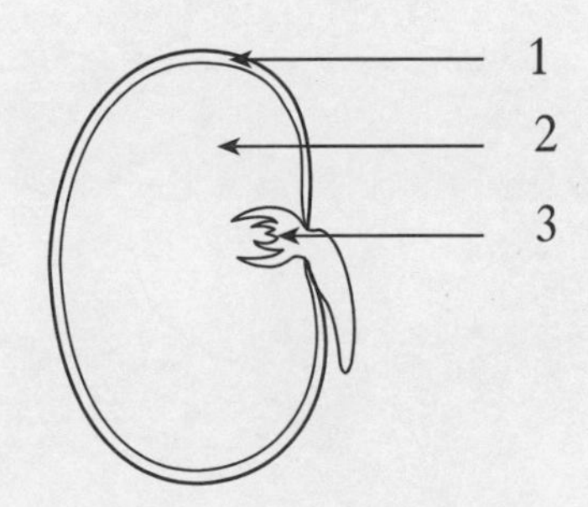

17. Which of the following correctly labels the parts of the seed?

	Part labelled		
	1	*2*	*3*
A	embryo plant	seed coat	food store
B	seed coat	food store	embryo plant
C	food store	embryo plant	seed coat
D	seed coat	embryo plant	food store

18. The function of the seed coat is to provide

A energy

B protection

C food

D a suitable temperature.

Questions 19 and 20 refer to the investigation below.

The following dishes were set up to investigate different conditions affecting the germination of cress seeds.

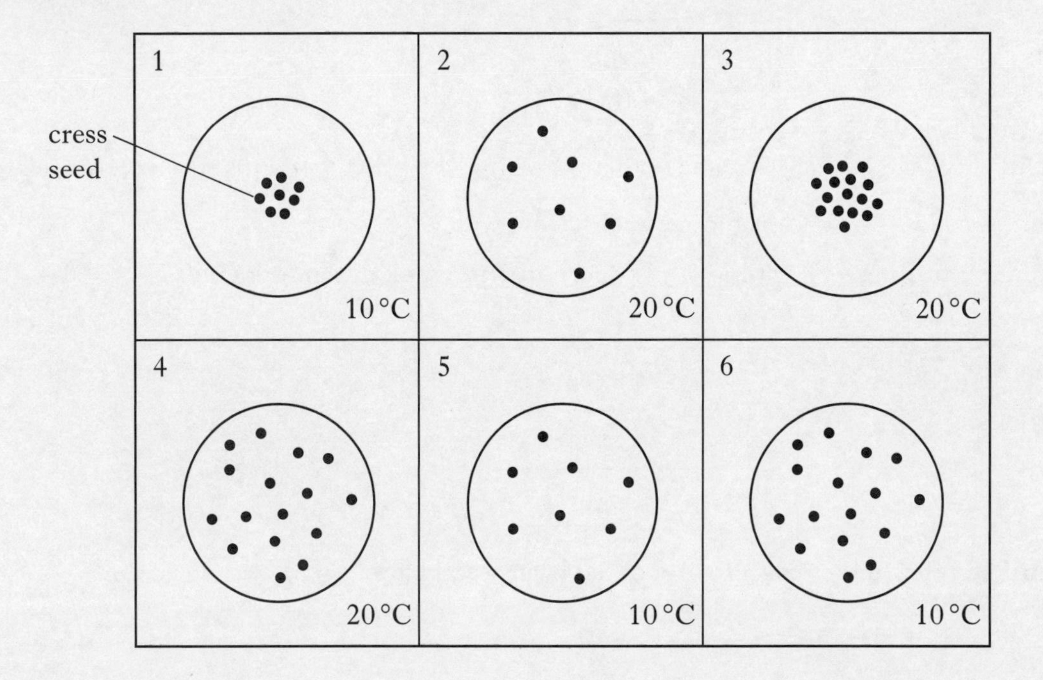

19. Which **two** dishes should be compared to find the effect of temperature on germination?

A 1 and 2

B 1 and 3

C 3 and 5

D 4 and 6

20. To investigate the effect of spacing on germination, dish 3 is best compared with

A dish 2

B dish 4

C dish 5

D dish 6.

21. The ratio of peat to sand in a compost is **3 : 1**.

Which of the following pie charts presents this information correctly?

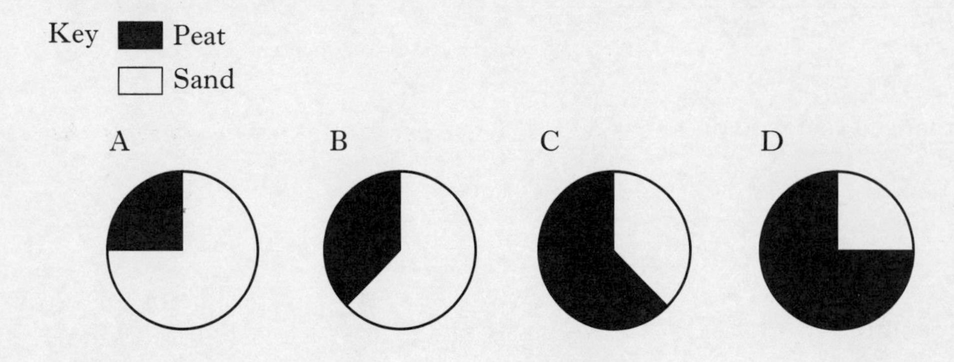

22. Which of the following **both** improve drainage when added to soil?

A Loam and peat

B Peat and sand

C Sand and perlite

D Loam and perlite

23. Which line in the table correctly matches the mineral to the part of the plant it encourages to grow?

	Mineral	
	Nitrogen (*N*)	*Phosphorus* (*P*)
A	leaf	root
B	root	leaf
C	leaf	leaf
D	root	root

24. The plant in the diagram needs to be

A dead headed

B watered

C potted on

D given fertiliser.

25. Grey mould in plants can be controlled by

A crushing

B insecticide

C fungicide

D pesticide.

Candidates are reminded that the answer sheet for Section A MUST be returned INSIDE this answer book.

[Turn over for Section B on *Page thirteen*

[BLANK PAGE]

DO NOT WRITE IN THIS MARGIN

SECTION B

All questions in this Section should be attempted.

Marks

1. Read the following passage carefully.

Inhalers can be used in the management of asthma as shown in the photograph.

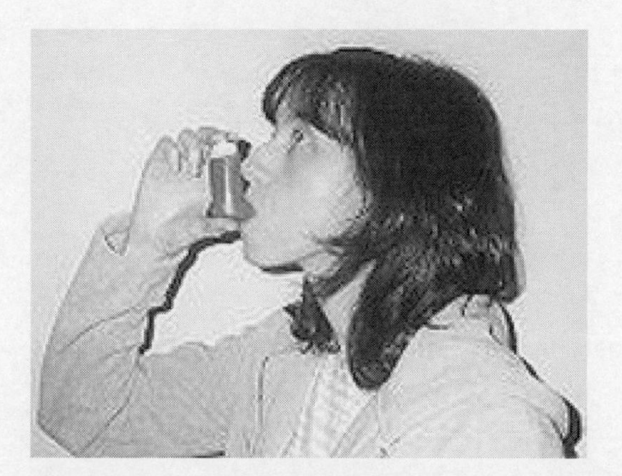

Asthma is a very common condition. Britain has one of the highest rates for asthma in the world, along with New Zealand, Australia and Ireland. In 2001 the number of people suffering from asthma in Britain was higher than ever before.

It was estimated that 4.5 million people – 1 in 13 adults and 1 in 8 children – were being treated for asthma in Britain in that year. In 1999 the estimated figure was 3.4 million.

Nobody knows for sure why asthma is becoming more common, but it is thought to be due to a complex combination of genetic and environmental factors.

The number of new cases of asthma each year is now three to four times higher in adults and six times higher in children than it was twenty five years ago.

Answer the questions below, using the information from the passage.

(*a*) Which four countries have the highest rates for asthma in the world?

____

____ **1**

(*b*) How many people were treated for asthma in Britain in 2001?

______________ million **1**

(*c*) Was the rate of asthma higher in adults or in children in 2001?

______________________________ **1**

(*d*) Which combination of factors is thought to have caused asthma to become more common?

____ **1**

[Turn over

DO NOT WRITE IN THIS MARGIN

Marks

2. (*a*) Forty adults were asked to name their favourite drink.
The results are shown below.

Favourite drink	*Number of adults*
Beer	20
Lemonade	10
Wine	5
Water	5

(i) What was the most popular type of drink?

____________________ **1**

(ii) Calculate the percentage of adults whose favourite drink was lemonade.
Space for calculation

____________ % **1**

(iii) Present the information in the table in the form of a pie chart.
(An additional pie chart, if required, will be found on page 27.)

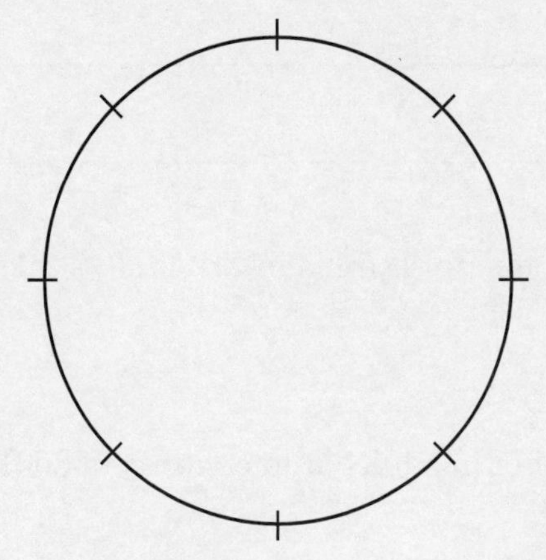

2

(*b*) State **one long-term** effect of drinking alcohol to excess.

____ **1**

DO NOT WRITE IN THIS MARGIN

Marks

3. (*a*) Body fat can be measured using a body fat sensor as shown in the photograph. The sensor is an example of a high tech instrument.

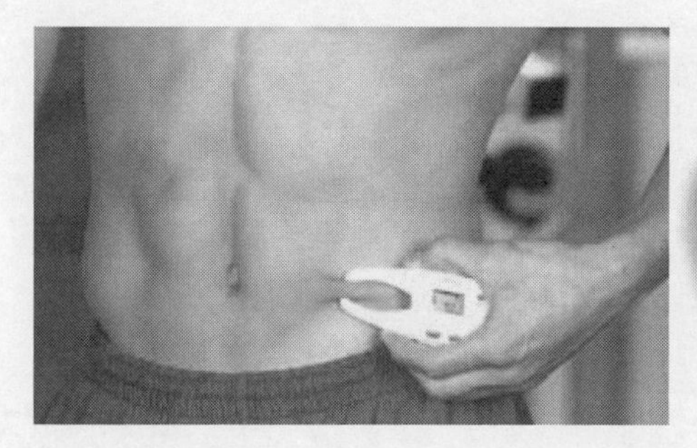

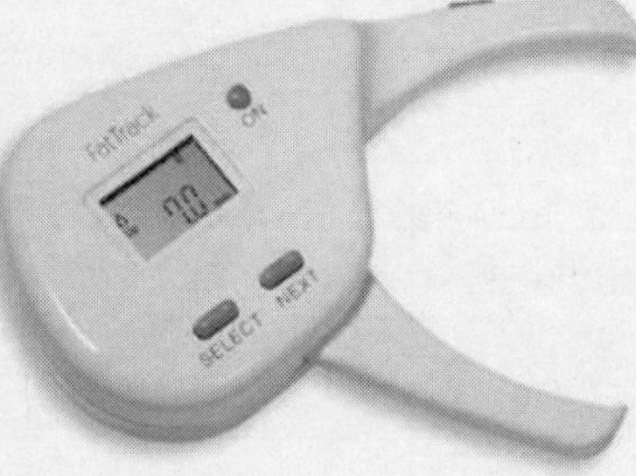

(i) Complete the table below by stating the measurement taken by each instrument and whether it is a high or low tech instrument.

Instrument	*Measurement*	*High or low tech*
Body fat sensor	Body fat	High
Clinical thermometer		
Digital sphygmomanometer		
Pulsometer		

2

(ii) State **one disadvantage** of using high tech instruments.

____ **1**

(*b*) Name **one** health condition which might occur as a result of being very overweight.

____ **1**

[Turn over

DO NOT WRITE IN THIS MARGIN

Marks

4. (*a*) The table below shows the percentage of men in different age groups who are light, medium and heavy smokers.

	Percentage of men					
	16–24 years	*25–34 years*	*35–44 years*	*45–54 years*	*55–64 years*	*65–74 years*
Light smokers	13	7	6	3	2	4
Medium smokers	19	17	11	13	13	8
Heavy smokers	6	15	18	17	17	8

(i) What percentage of men aged 45–54 years are heavy smokers?

_______________ % **1**

(ii) Calculate the percentage of men aged 16–24 years who **do not** smoke.

Space for calculation

_______________ % **1**

DO NOT WRITE IN THIS MARGIN

Marks

4. (*a*) (continued)

(iii) On the grid below, complete the **bar graph** by

(1) putting a label on the vertical axis **1**

(2) plotting the remaining results for heavy smokers. **1**

(Additional graph paper, if required, will be found on page 27.)

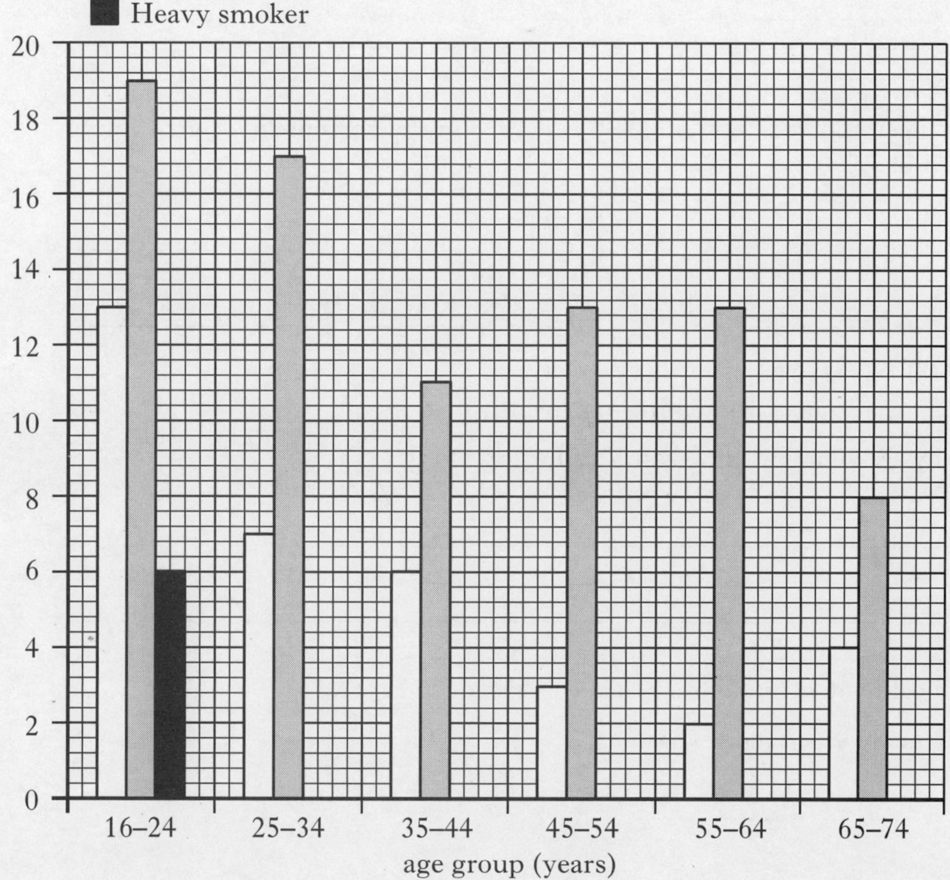

(iv) Draw **one** conclusion from the table.

____

____ **1**

(*b*) Name **one** substance present in cigarette smoke which reduces the ability of the blood to carry oxygen.

______________________________ **1**

[Turn over

DO NOT WRITE IN THIS MARGIN

Marks

5. The table below shows the percentage of 11–15 year olds who ate fresh fruit daily.

	11–15 year olds who ate fresh fruit daily (%)	
Year	*Boys*	*Girls*
1990	49·4	60·4
1994	60·9	69·0
1998	62·4	68·4

(*a*) What change occurred between 1990 and 1998 in the percentage of girls eating fresh fruit daily?

____

____ **1**

(*b*) Fresh fruit is a source of vitamins in a healthy diet.

(i) Why are vitamins important in a healthy diet?

____

____ **1**

(ii) A healthy diet contains a balance of three food types.

Name the three food types.

1 ____________________

2 ____________________

3 ____________________ **1**

DO NOT WRITE IN THIS MARGIN

Marks

6. The growth of sunflowers from seeds of different varieties was investigated by five students.

Each student grew a single seed from one of the varieties in the **same conditions**. After eight weeks, the height of each plant was measured.

The results are shown in the table below.

Variety of sunflower	*Height after eight weeks* (cm)
Teddy Bear	40
Pacino	0
Ring of Fire	100
Sunspot	45
Vanilla Ice	150

(*a*) (i) Which variety of sunflower grew fastest?

________________________ **1**

(ii) Suggest **one** improvement to make the results more **reliable**.

________________________ **1**

(*b*) A suitable temperature is required to allow the sunflower seeds to germinate. Name **one** other condition required for seed germination.

________________________ **1**

(*c*) Some seeds will not germinate until spring when the soil temperature rises. What is the name given to this delay in germination?

________________________ **1**

(*d*) After a seed has germinated the plant makes its own food using sunlight. Name this process of food production.

________________________ **1**

[Turn over

DO NOT WRITE IN THIS MARGIN

7. (*a*) A student carried out the following steps to take a cutting from a *Geranium* plant.

Step 1 A diagonal cut was made in the stem.

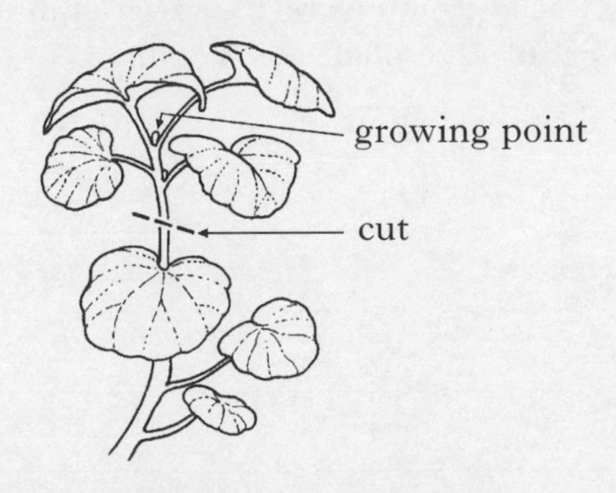

Step 2 The lower leaves of the cutting were removed.

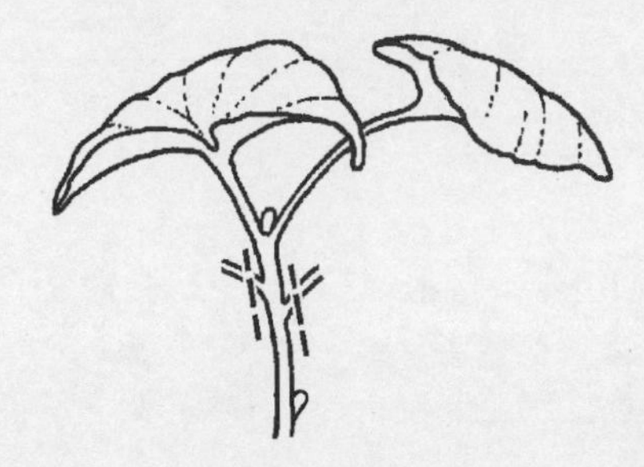

Step 3 A hole was made in the centre of the compost in a plant pot.

Step 4 The cutting was placed in the hole and secured by pressing down gently on the compost.

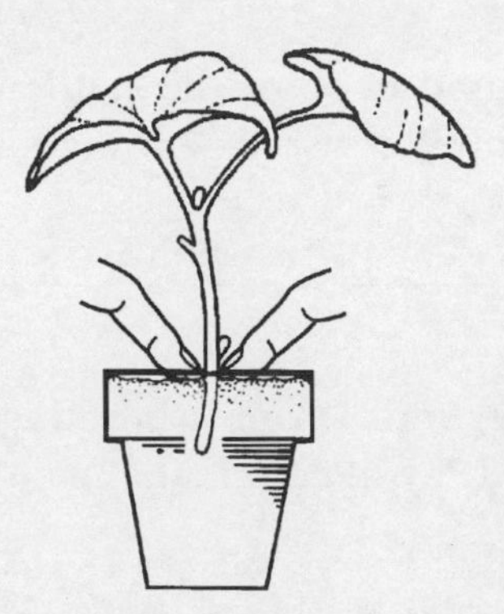

Step 5 The compost was dampened by adding water.

Step 6 The potted cutting was then placed in a mist propagator.

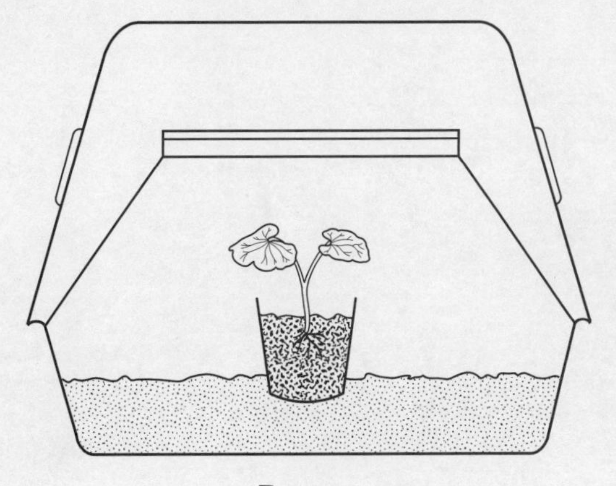

DO NOT WRITE IN THIS MARGIN

Marks

7. (*a*) (continued)

(i) What name is given to the growing point shown in **Step 1**?

______________________________ **1**

(ii) Describe an additional step that could be taken to encourage root growth.

____ **1**

(iii) Suggest a different method of increasing humidity if a mist propagator was not available for **Step 6**.

____ **1**

(*b*) Another method of propagating plants involves the pegging down of stems into soil. This is shown in the diagram below.

(i) What name is given to this method of pegging down?

______________________________ **1**

(ii) State **one** advantage of this method.

____ **1**

[Turn over

DO NOT WRITE IN THIS MARGIN

Marks

8. (*a*) The temperature in a greenhouse was recorded once each day for a week.

The results are shown in the table below.

Day of week	*Temperature* (°C)
Monday	24
Tuesday	22
Wednesday	23
Thursday	24
Friday	20
Saturday	13
Sunday	21

(i) Calculate the **average** temperature for the week.

Space for calculation

________ °C **1**

(ii) Describe a possible source of error that could have given the result obtained on Saturday.

____

______________________________ **1**

(*b*) During warm, sunny weather the temperature in a greenhouse can rise so high that the plants can be damaged.

State **one** way of reducing the temperature in a greenhouse.

______________________________ **1**

DO NOT WRITE IN THIS MARGIN

Marks

9. Seeds can be sown on capillary matting or in pots. These methods of sowing seeds were compared using three types of Cyclamen: *C. balearicum*, *C. coum* and *C. libanoticum*.

Ten seeds of each type were sown in pots and fifteen seeds of each type were sown on capillary matting.

C. balearicum took 89 days to germinate in pots and 120 days on capillary matting. *C. coum* took 51 days to germinate in pots and 55 days on capillary matting. *C. libanoticum* took 57 days to germinate in pots and 62 days on capillary matting.

(*a*) Use this information to complete the table by

(i) providing headings **1**

(ii) putting in the results for each type. **1**

Type of Cyclamen	*Time to germinate (days)*	
C. balearicum		
C. coum		
C. libanoticum		

(*b*) Which type of Cyclamen was the slowest to germinate?

____________________________ **1**

(*c*) Suggest **one** way in which the experimental procedure could be improved to make the comparison **valid**.

____ **1**

[Turn over

DO NOT WRITE IN THIS MARGIN

Marks

10. A student carried out an investigation to compare how well a washing powder worked at different temperatures.

The results are shown in the table below.

Temperature (°C)	*Percentage of stain removed*
10	80
20	88
30	94
40	92
50	82
60	63

(*a*) On the grid below complete the **line graph** by

(i) providing a label for the horizontal axis **1**

(ii) completing the scale on the vertical axis **1**

(iii) plotting the remaining results. **1**

(Additional graph paper, if required, will be found on page 28.)

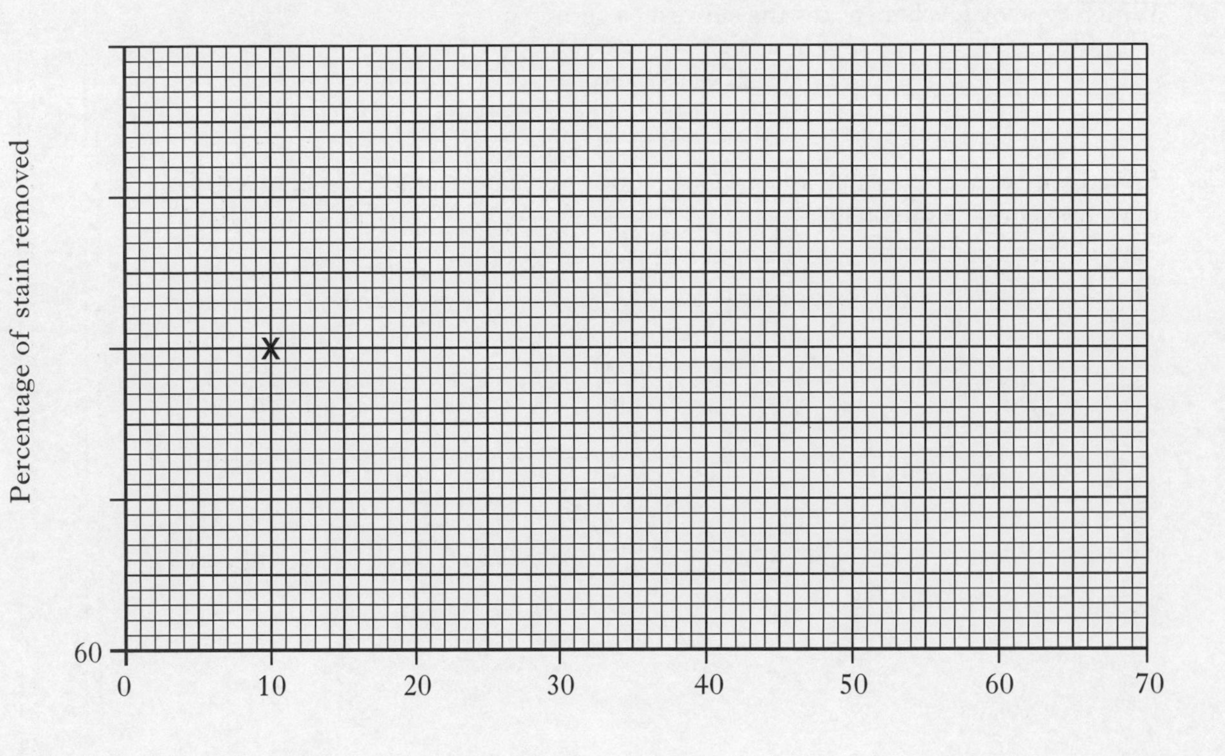

DO NOT WRITE IN THIS MARGIN

Marks

10. (continued)

(*b*) The diagram below represents the temperature dial on a washing machine.

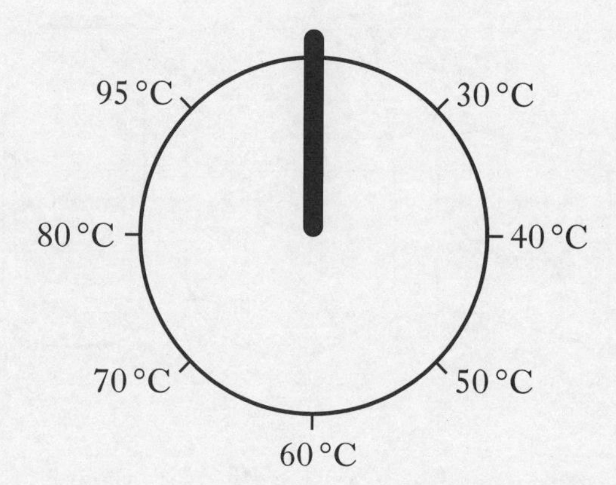

Using the information in the table, **circle** the temperature that should be recommended to get the best results from this washing powder. **1**

(*c*) **Biological** washing powders contain enzymes.

(i) Name the organisms used to produce these enzymes.

______________________________ **1**

(ii) The enzymes are enclosed in a harmless coating.
Suggest a reason for this.

____ **1**

(iii) Biological washing powders are used at lower temperatures and this reduces fuel consumption.
Explain why this can reduce pollution in the environment.

____

____ **1**

[Turn over

DO NOT WRITE IN THIS MARGIN

Marks

11. (*a*) Four types of milk are shown in the diagram below.

Complete the table below to match the type of milk to its treatment.

Treatment	*Type of milk*
Heat treated to kill all bacteria giving it a longer shelf life	
Nearly all the fat is removed	
Half the water is removed	
About half of the fat is removed	

2

(*b*) Before the milk reaches the supermarket, samples are tested with resazurin.
What does the resazurin test detect in the milk samples?

____

1

(*c*) **<u>Underline</u> one** option in each set of brackets to make the sentences below correct.

Rennet is added to milk during cheese-making to clot {protein. / fat.}

The liquid left over after clotting is called {curds. / whey.}

1

[*END OF QUESTION PAPER*]

SPACE FOR ANSWERS

DO NOT WRITE IN THIS MARGIN

ADDITIONAL PIE CHART FOR QUESTION 2(*a*)(iii)

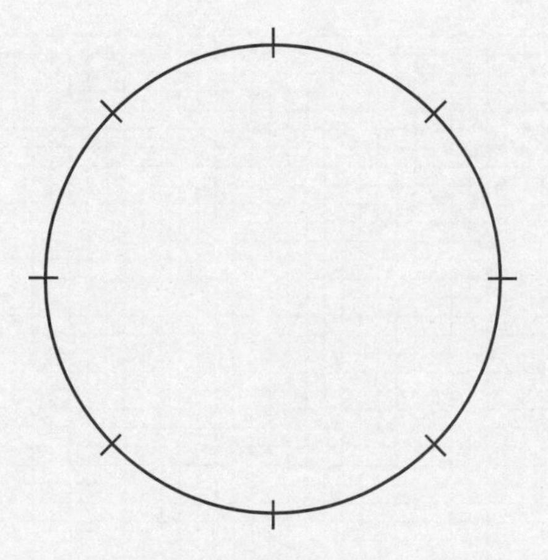

ADDITIONAL GRAPH PAPER FOR QUESTION 4(*a*)(iii)

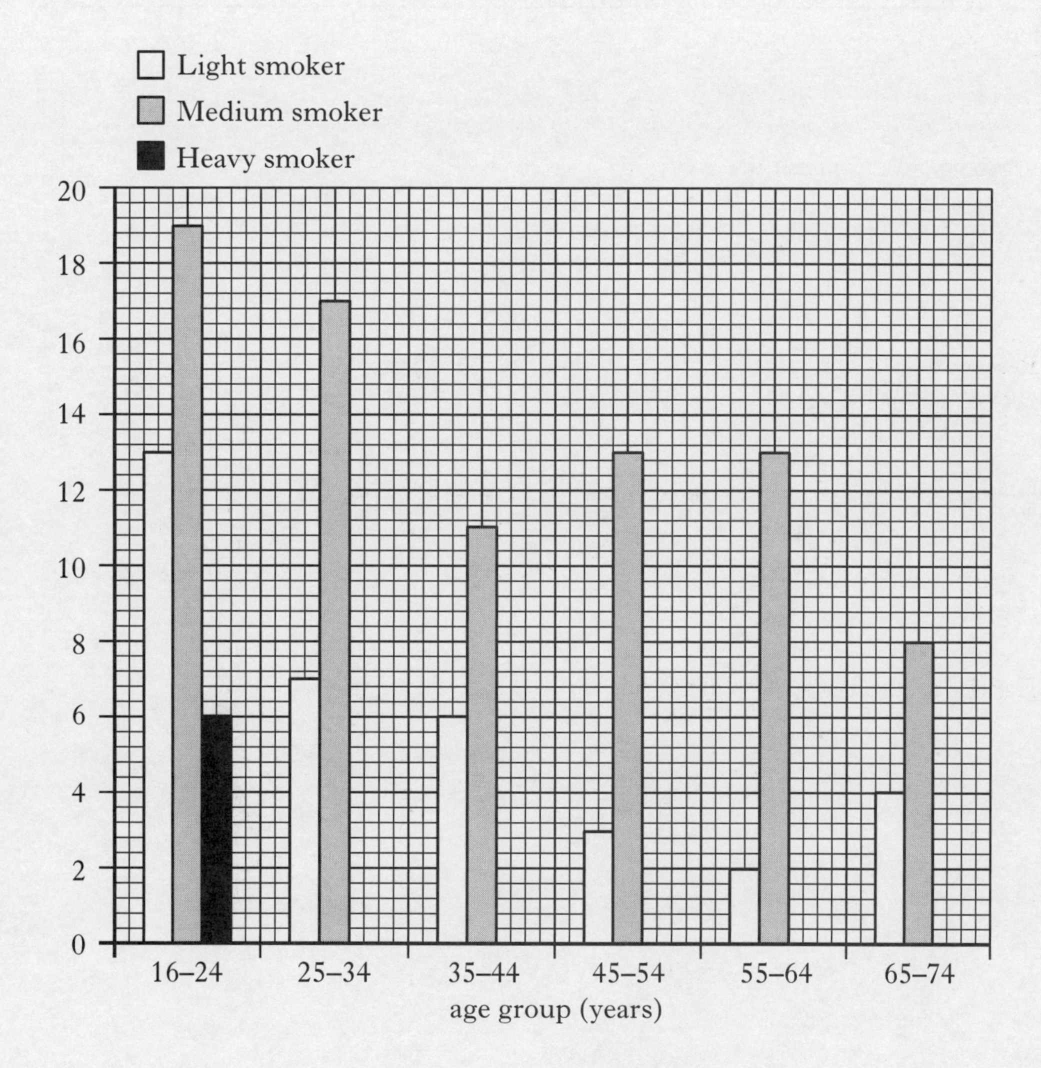

SPACE FOR ANSWERS

DO NOT WRITE IN THIS MARGIN

ADDITIONAL GRAPH PAPER FOR QUESTION 10(*a*)

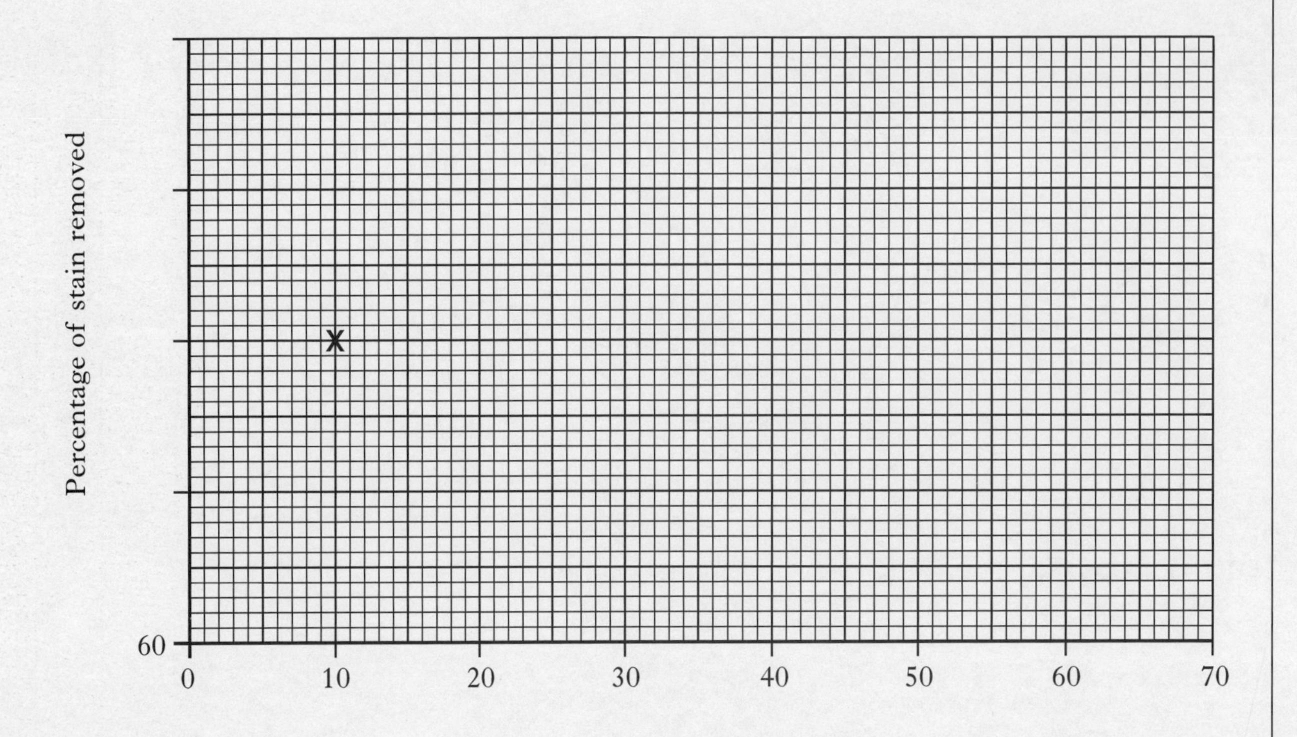

[BLANK PAGE]

FOR OFFICIAL USE

Section B Total

X007/101

NATIONAL QUALIFICATIONS 2005

WEDNESDAY, 18 MAY
9.00 AM – 10.30 AM

BIOLOGY INTERMEDIATE 1

Fill in these boxes and read what is printed below.

Full name of centre

Town

Forename(s)

Surname

Date of birth
Day Month Year

Scottish candidate number

Number of seat

SECTION A

Instructions for completion of Section A are given on page two.

SECTION B

1 All questions should be attempted.
2 The questions may be answered in any order but all answers are to be written in the spaces provided in this answer book, and must be written clearly and legibly in ink.
3 Additional space for answers will be found at the end of the book. If further space is required, supplementary sheets may be obtained from the invigilator and should be inserted inside the **front** cover of this book.
4 The numbers of questions must be clearly inserted with any answers written in the additional space.
5 Rough work, if any should be necessary, should be written in this book and then scored through when the fair copy has been written. If further space is required, a supplementary sheet for rough work may be obtained from the invigilator.
6 Before leaving the examination room you must give this book to the invigilator. If you do not, you may lose all the marks for this paper.

SCOTTISH QUALIFICATIONS AUTHORITY

Read carefully

1 Check that the answer sheet provided is for **Biology Intermediate 1 (Section A)**.

2 Check that the answer sheet you have been given has **your name**, **date of birth**, **SCN** (Scottish Candidate Number) and **Centre Name** printed on it.

Do not change any of these details.

3 If any of this information is wrong, tell the Invigilator immediately.

4 If this information is correct, **print** your name and seat number in the boxes provided.

5 Use **black** or **blue ink** for your answers. **Do not use red ink.**

6 The answer to each question is **either** A, B, C or D. Decide what your answer is, then put a horizontal line in the space provided (see sample question below).

7 There is **only one correct** answer to each question.

8 Any rough working should be done on the question paper or the rough working sheet, **not** on your answer sheet.

9 At the end of the exam, put the **answer sheet for Section A inside the front cover of this answer book**.

Sample Question

Which of the following foods contains a high proportion of fat?

A Bread

B Butter

C Sugar

D Apple

The correct answer is **B**—Butter. The answer **B** has been clearly marked with a horizontal line (see below).

A B C D

Changing an answer

If you decide to change your answer, cancel your first answer by putting a cross through it (see below) and fill in the answer you want. The answer below has been changed to **B**.

A B C D

If you then decide to change back to an answer you have already scored out, put a tick (✓) to the **right** of the answer you want, as shown below:

A B ✓ C D or A B ✓ C D

SECTION A

All questions in this Section should be attempted.

Answers should be given on the separate answer sheet provided.

1. Which of the following shows the health triangle correctly?

A
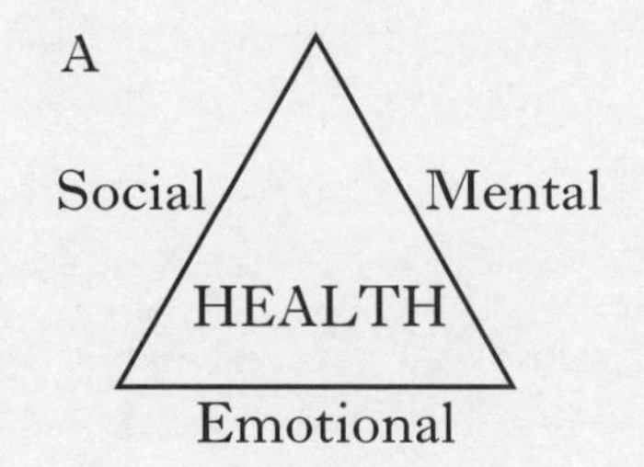

B
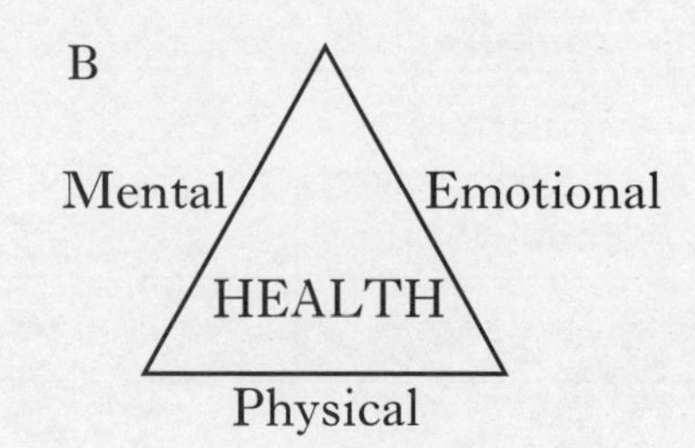

C
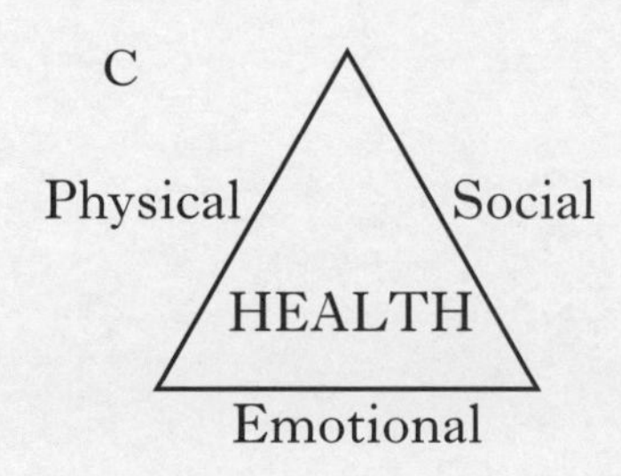

D
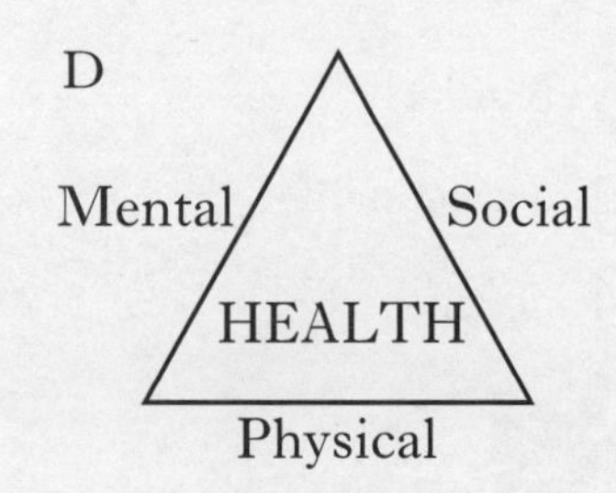

2. Which of the following is **most** likely to lead to high blood pressure?

	Diet	*Exercise*
A	high fat	regular
B	low fat	none
C	high fat	none
D	low fat	regular

[Turn over

3. A number of people were asked how often they exercised.

The results are shown in the table below.

How often exercise is taken	*Number of people*
regularly	70
occasionally	20
never	30

Which pie-chart represents this information correctly?

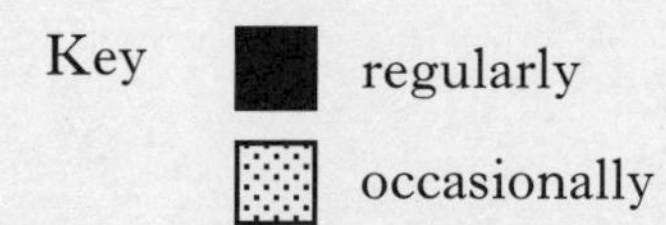

A

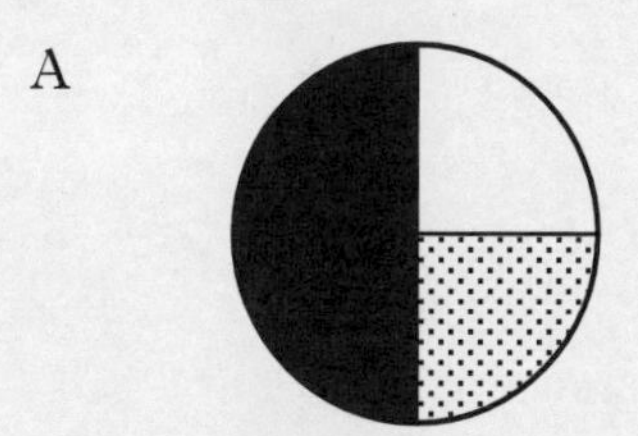

B

C

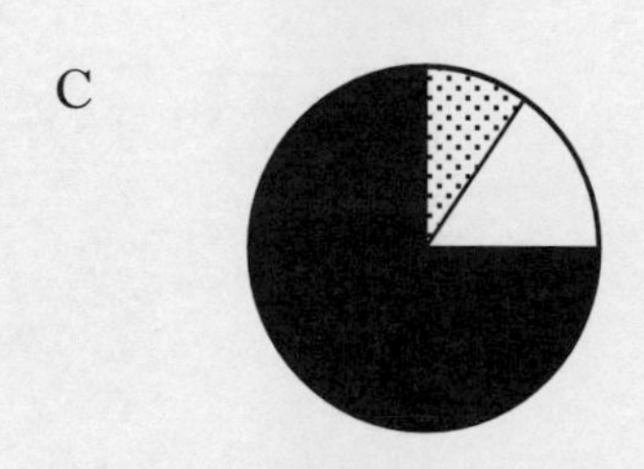

D

4. A student measured his pulse rate and found it to be 12 beats in a **10 second period**.

His heart rate, in **beats per minute**, was

A 12

B 60

C 72

D 120.

5. The table below shows the resting pulse rates and recovery times for four students.

Student	*Resting Pulse Rate* (beats per minute)	*Recovery Time* (minutes)
A	60	5
B	60	10
C	80	10
D	80	5

Which student is likely to be the fittest?

[Turn over

Questions 6 and 7 refer to the diagram of the human breathing system shown below.

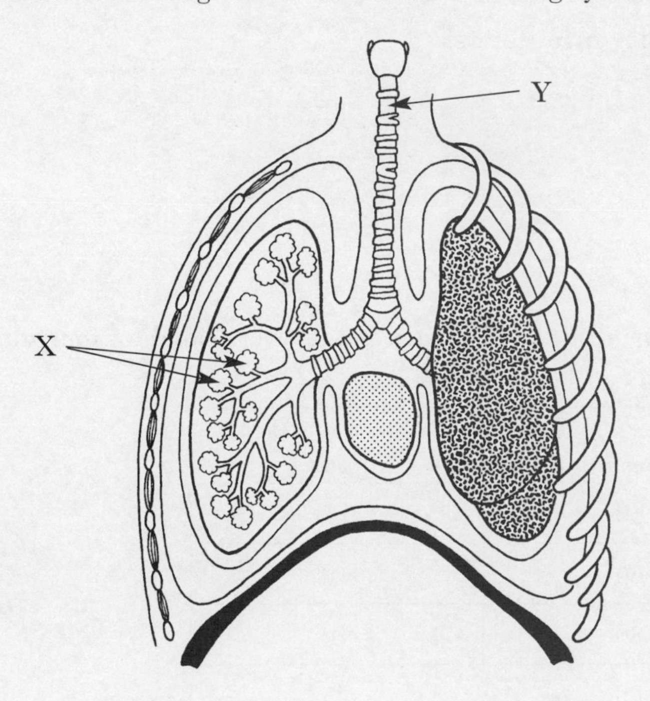

6. Which line in the table identifies correctly structures X and Y?

	X	*Y*
A	air sacs	bronchus
B	air sacs	windpipe
C	bronchioles	windpipe
D	bronchioles	bronchus

7. The function of the structures labelled X is to allow

A oxygen to enter and leave the blood

B carbon dioxide to enter and leave the blood

C carbon dioxide to enter and oxygen to leave the blood

D oxygen to enter and carbon dioxide to leave the blood.

8. The diagram below shows an instrument used by people with asthma.

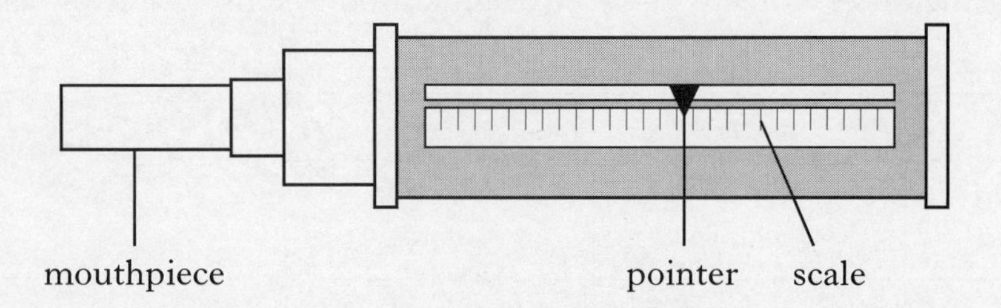

People with asthma use this instrument to measure

A peak flow

B tidal volume

C vital capacity

D breathing rate.

9. The table below shows the blood groups of fifty students.

Blood group	*Number of students*
A	8
B	10
AB	5
O	27

What percentage of the students have blood group AB?

A 5%

B 10%

C 23%

D 46%

10. What is the name of the test used to show if milk is safe to drink?

A Pasteurisation

B Preservation

C Resazurin

D Immobilisation

[Turn over

11. Fresh milk is treated to produce different types of milk.

Which line in the table identifies correctly the type of milk produced by each treatment?

	Treatment		
	Some liquid removed by heating	*Nearly all fat removed*	*Some fat removed*
A	semi-skimmed	skimmed	evaporated
B	evaporated	skimmed	semi-skimmed
C	skimmed	evaporated	semi-skimmed
D	skimmed	semi-skimmed	evaporated

12. The diagram below shows some information about the composition of milk.

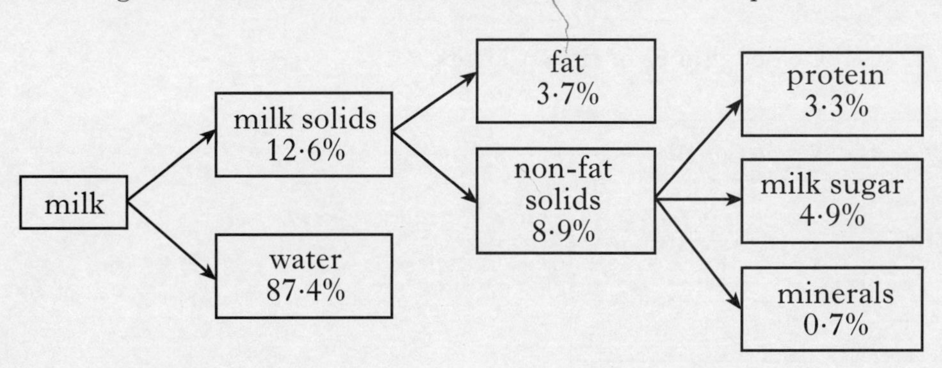

Which of the following conclusions can be drawn from this information?

A There are more milk solids than water in milk.

B Milk consists mainly of non-fat solids.

C Milk has a higher percentage of fat than sugar.

D There is seven times as much sugar as minerals in milk.

13. During yoghurt making, which of the following is converted into an acid?

A Sugar

B Protein

C Fat

D Rennet

14. The diagram shows an investigation on the production of cheese from milk.

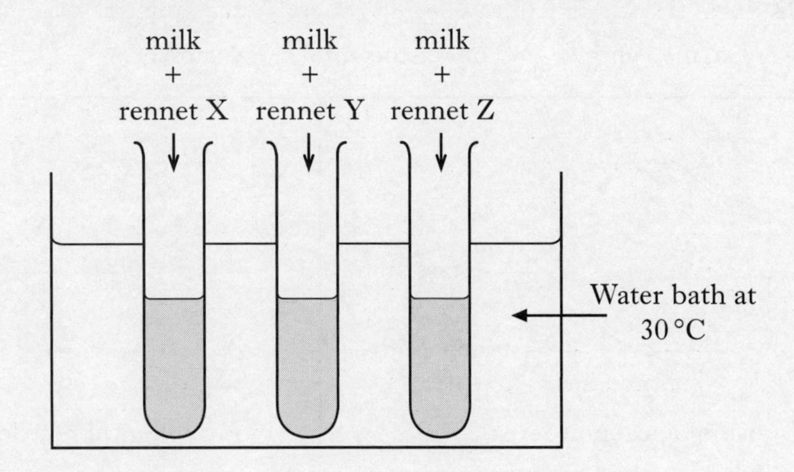

The table below shows the time taken for cheese to form.

Rennet	*Time taken for cheese to form (minutes)*
X	80
Y	50
Z	60

Which variable was altered in this investigation?

A Temperature

B Time

C Type of rennet

D Type of milk

15. Which of the following are **both** sources of rennet?

A Bacteria and genetically engineered fungi

B Calves and genetically engineered fungi

C Bacteria and calves

D Yeast and bacteria

[Turn over

Questions 16 and 17 refer to the following investigation into the effectiveness of washing powders on removing stains.

Six cloths were stained and washed under the conditions shown.

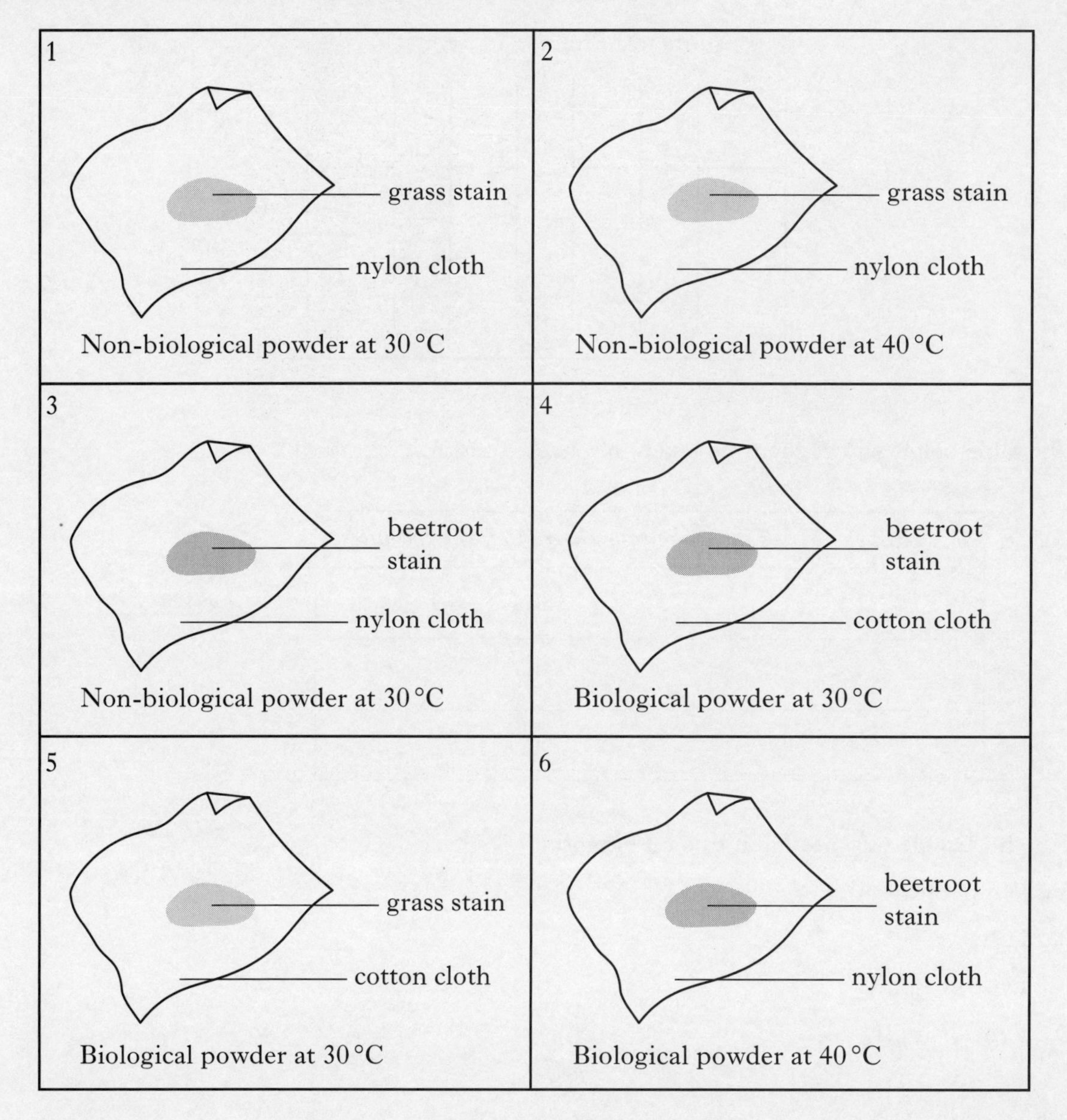

16. Which two cloths should be compared to investigate the effect of temperature on stain removal?

A 1 and 2

B 1 and 6

C 2 and 3

D 5 and 6

17. Which factor is being investigated by comparing cloths 4 and 5?

A Type of detergent

B Type of stain

C Temperature

D Type of cloth

18. The picture shows mist propagation of cuttings growing in a greenhouse.

The main reasons for growing cuttings in this way are to

A decrease humidity and increase temperature

B decrease humidity and decrease temperature

C increase humidity and increase temperature

D increase humidity and decrease temperature.

19. Which of the following would **increase** water retention when added to soil?

A Peat

B Grit

C Sand

D Perlite

20. Which of the following are both methods of controlling aphids?

A Soapy water and insecticides

B Insecticides and dead heading

C Dead heading and fungicide

D Fungicide and soapy water

[Turn over

Questions 21 and 22 refer to the following diagram of a seed.

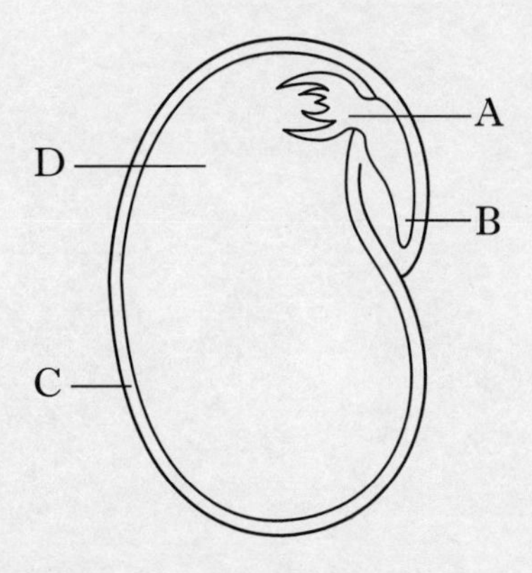

21. Which part is the food store?

22. The function of part C is to

A grow into new root

B grow into new shoot

C protect the seed

D provide energy for growth.

23. What name is given to fine seeds that are enclosed in a ball of clay?

A Chitted

B Pre-germinated

C Pelleted

D Non-pelleted

24. A student investigated if water was needed for cress seeds to germinate.

The diagram below shows a dish she set up.

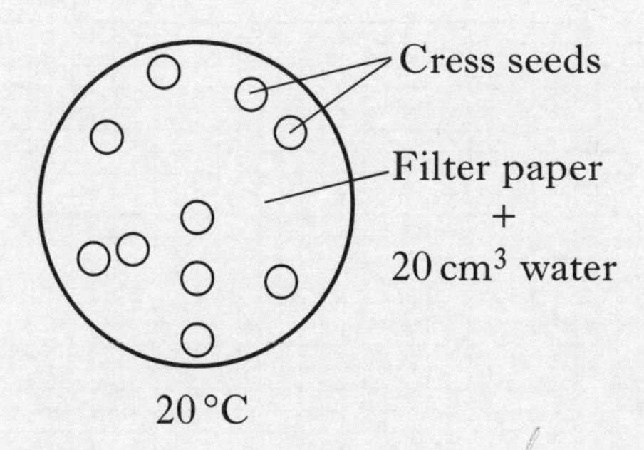

Which of the following dishes would be a suitable **control** for this experiment?

A

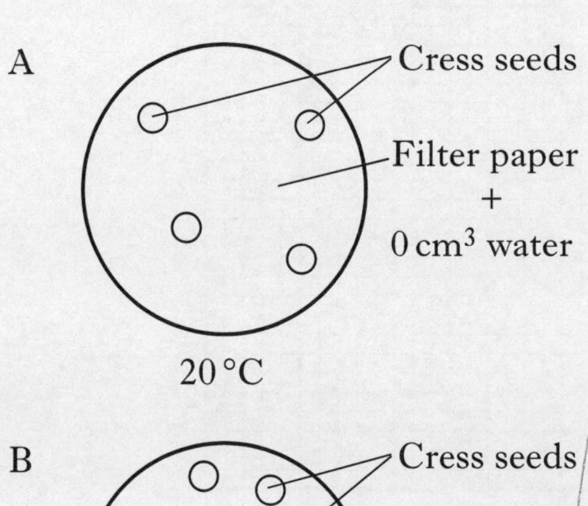

B

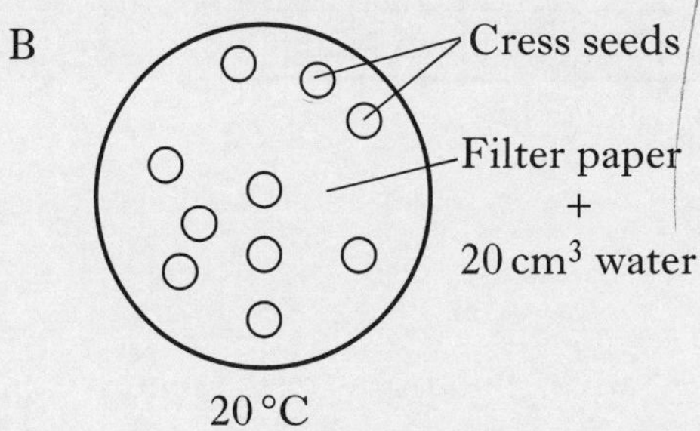

C

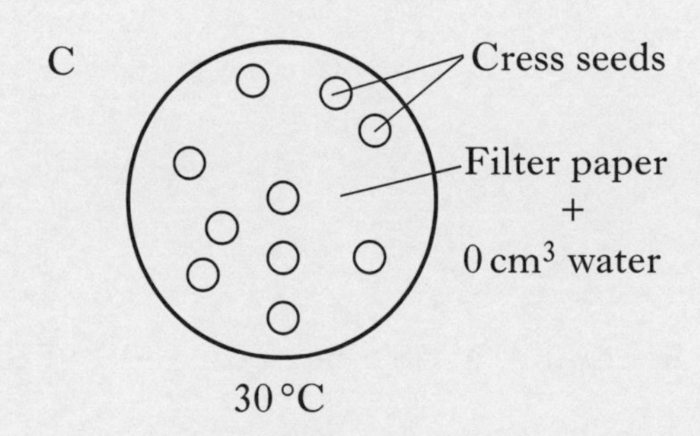

D

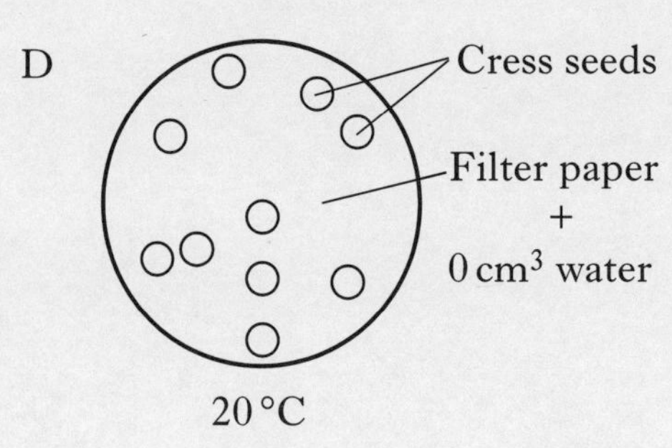

[Turn over

25. The graph below shows the results of an investigation into the effect of temperature on the rate of photosynthesis.

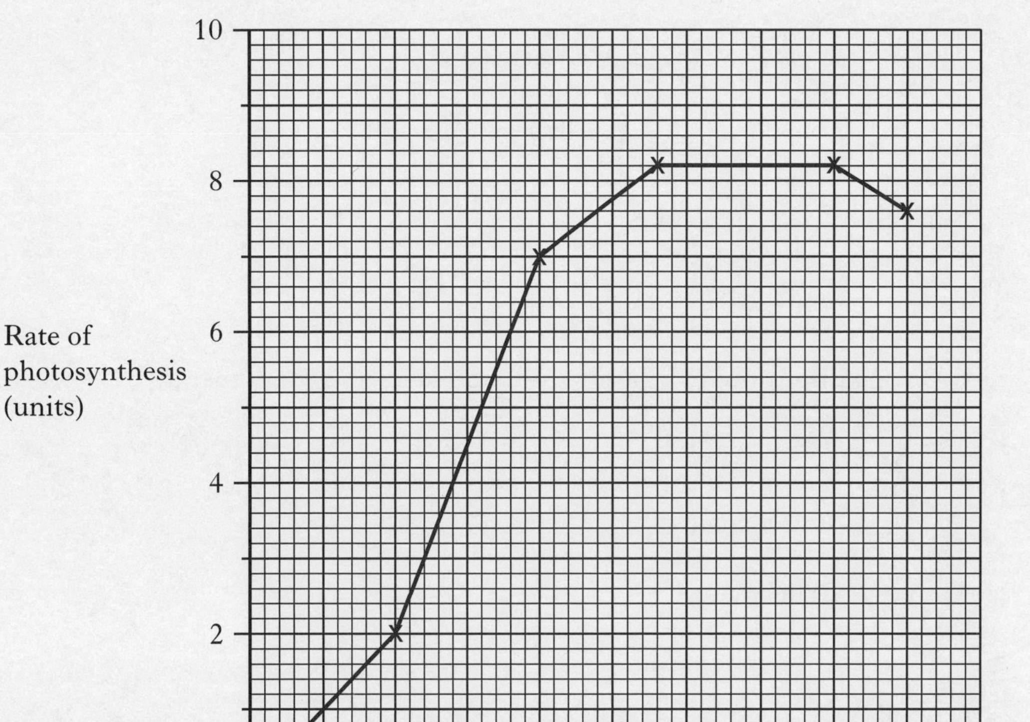

Between which two temperatures is the **increase** in the rate of photosynthesis greatest?

A 0 – 2 °C

B 0 – 10 °C

C 2 – 7°C

D 10 – 20 °C

Candidates are reminded that the answer sheet for Section A MUST be returned INSIDE the front cover of this answer book.

DO NOT WRITE IN THIS MARGIN

SECTION B

All questions in this Section should be attempted.

Marks

1. The diagram below represents the human heart. The arrows indicate the direction of blood flow through the blood vessels.

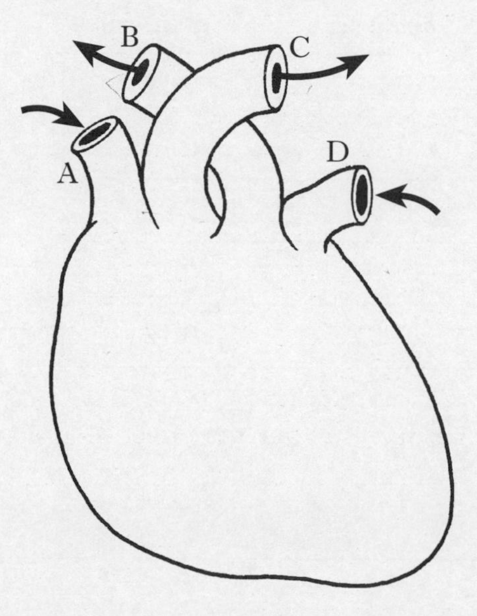

(*a*) What is the function of the heart?

____ **1**

(*b*) Which **two** letters in the diagram identify arteries?

Letters ________ and ________ **1**

(*c*) Name the type of blood vessels which link arteries to veins.

____ **1**

[Turn over

DO NOT WRITE IN THIS MARGIN

Marks

2. (*a*) The reaction time of five students was measured using an electronic timer.

The results are shown in the table below.

Student	*Reaction time (s)*		
	First attempt	*Second attempt*	*Third attempt*
1	0·35	0·24	0·19
2	0·27	0·20	0·14
3	0·15	0·10	0·08
4	0·27	0·19	0·15
5	0·24	0·17	0·12

(i) What conclusion can be drawn about the effect of practice on reaction time?

____

____ **1**

(ii) Which student had the fastest reaction time?

Student ______________ **1**

(*b*) State **one** factor, other than practice, which can affect reaction time.

____ **1**

DO NOT WRITE IN THIS MARGIN

Marks

3. A healthy, varied diet should provide all the vitamins a person needs.

The table shows information about some vitamins.

Vitamin	*Good food sources*	*Importance*
A	liver, eggs, butter, margarine, oily fish	healthy eyes and skin
B_1	pork, breakfast cereal, wholemeal bread	releasing energy from food
B_{12}	red meat, liver, chicken, eggs	healthy nervous system
C	oranges, strawberries, kiwi fruit, potatoes	healthy immune system
D	eggs, cheese, oily fish	healthy bones and teeth
E	butter, margarine, avocado, muesli, olive oil	healthy skin
K	broccoli, cabbage, yoghurt	blood clotting and healthy bones and teeth

Use the information in the table to answer the following questions.

(*a*) Which **two** vitamins are important for healthy bones and teeth?

__________ and __________ **1**

(*b*) Name **one** food which is a good source of vitamins A, B_{12} and D.

____________________ **1**

(*c*) Which vitamins are found in margarine and butter?

____________________ **1**

(*d*) Suggest why an athlete might need a lot of vitamin B_1.

____

____ **1**

[Turn over

DO NOT WRITE IN THIS MARGIN

Marks

4. The table shows the percentage of the food groups found in rice.

Food Group	*Percentage* (%)
carbohydrate	80
fat	5
protein	10
other	5

(*a*) Use the information in the table to complete and label the pie chart below.

(An additional pie chart, if required, will be found on page 33.)

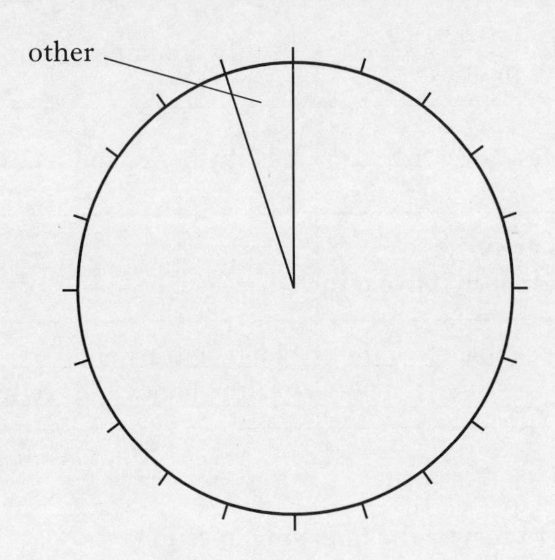

2

(*b*) Calculate the ratio of protein to fat in rice.

Show your answer as a simple whole number ratio.

Space for calculation

________ : ________
Protein Fat

1

DO NOT WRITE IN THIS MARGIN

Marks

5. (*a*) Draw lines to join each of the physiological measurements to the correct instrument.

Physiological measurement	*Instrument*
temperature	finger and stopwatch
body fat	clinical thermometer
blood pressure	skin fold callipers
heart rate	digital sphygmomanometer

2

(*b*) Physiological measurements can be made using high or low tech instruments.

(i) Name the high tech instrument in the list above.

______________________________________ **1**

(ii) State **one advantage** of using a high tech instrument.

____

______________________________________ **1**

[Turn over

DO NOT WRITE IN THIS MARGIN

Marks

6. Read the following passage carefully.

Warning on drugs in river water

Adapted from *The Guardian* newspaper

Waste Water Treatment Plant

Antibiotics and other medicines given to humans and livestock are contaminating European rivers.

High concentrations of antibiotics have been found in hospital and household sewage. Antibiotics can also reach the environment directly from the urine and faeces of farm animals.

Twenty-five pharmaceutical compounds, including five antibiotics, have been found in German rivers. Similar high levels of contamination have been found in other European rivers.

Scientists worry that the increasing levels of antibiotics may damage the environment and lead to an increase in antibiotic resistance.

New ways of reducing the contamination of waste water are needed. One suggested method is to collect the urine of hospital patients and use separation techniques to remove the antibiotics.

Use information from the passage to answer the questions below.

(*a*) Name **one** type of sewage in which antibiotics have been found.

____________________________ **1**

(*b*) State why the increasing levels of antibiotics worry scientists.

____

____ **1**

DO NOT WRITE IN THIS MARGIN

Marks

6. (continued)

(*c*) State **one** way in which contamination of waste water could be reduced.

____

____ 1

(*d*) What percentage of the pharmaceutical compounds found in German rivers was antibiotics?

Space for calculation

_______% 1

7. (*a*) **Tick the correct box** for each statement in the table to show whether it refers to antibiotics or antifungals.

Statement	*Antibiotics*	*Antifungals*
Used to treat athlete's foot and thrush		
Act on bacteria but not viruses		
Limit growth of fungi		
Produced naturally by soil fungi		

2

(*b*) Name the type of vessel in which antibiotics can be commercially produced.

______________________ 1

[Turn over

DO NOT WRITE IN THIS MARGIN

8. A student carried out an investigation into the effect of temperature on the reproduction of yeast cells.

Four identical flasks were set up and placed in water baths at different temperatures as shown below.

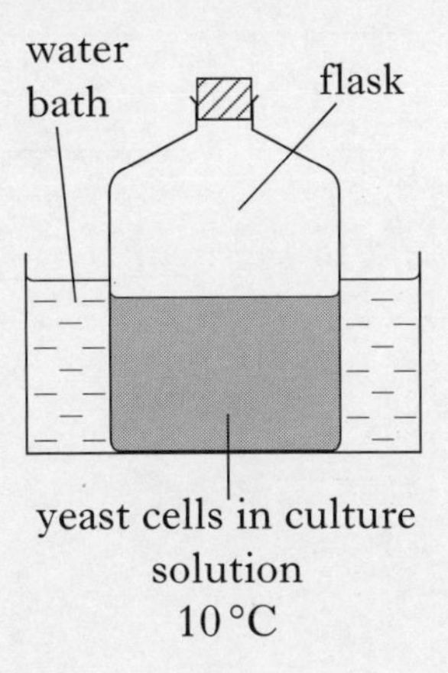

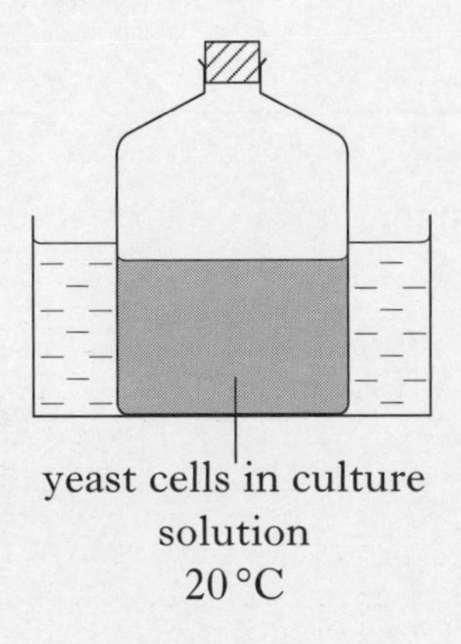

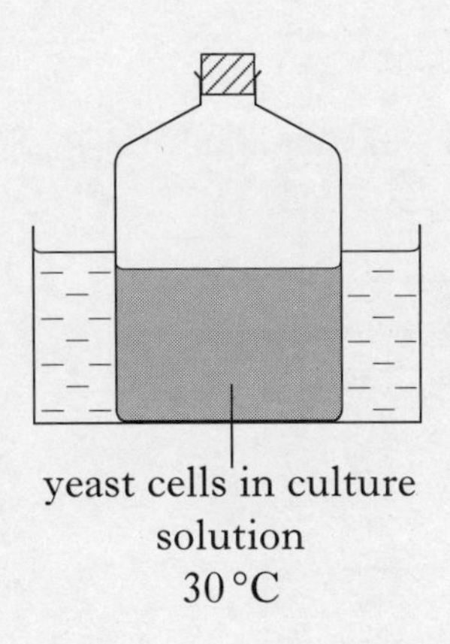

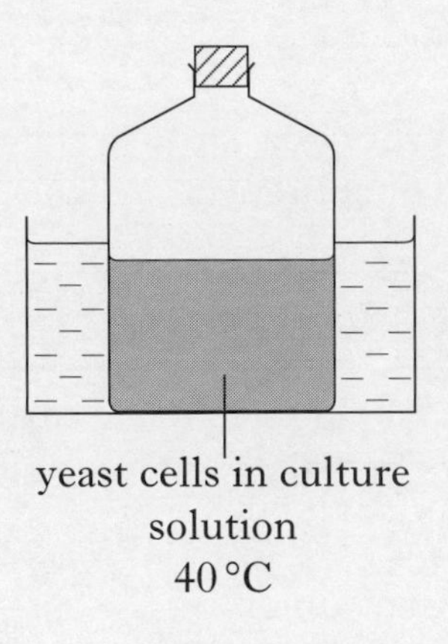

The results are shown in the table.

Temperature (°C)	*Number of yeast cells at start* (per mm^3)	*Number of yeast cells after 2 hours* (per mm^3)
10	50	60
20	55	100
30	50	560
40	45	140

DO NOT WRITE IN THIS MARGIN

Marks

8. (continued)

(*a*) On the grid complete the **line graph** for the number of yeast cells after 2 hours by

(i) putting a scale on the vertical axis **1**

(ii) plotting the graph. **1**

(Additional graph paper, if required, will be found on page 33.)

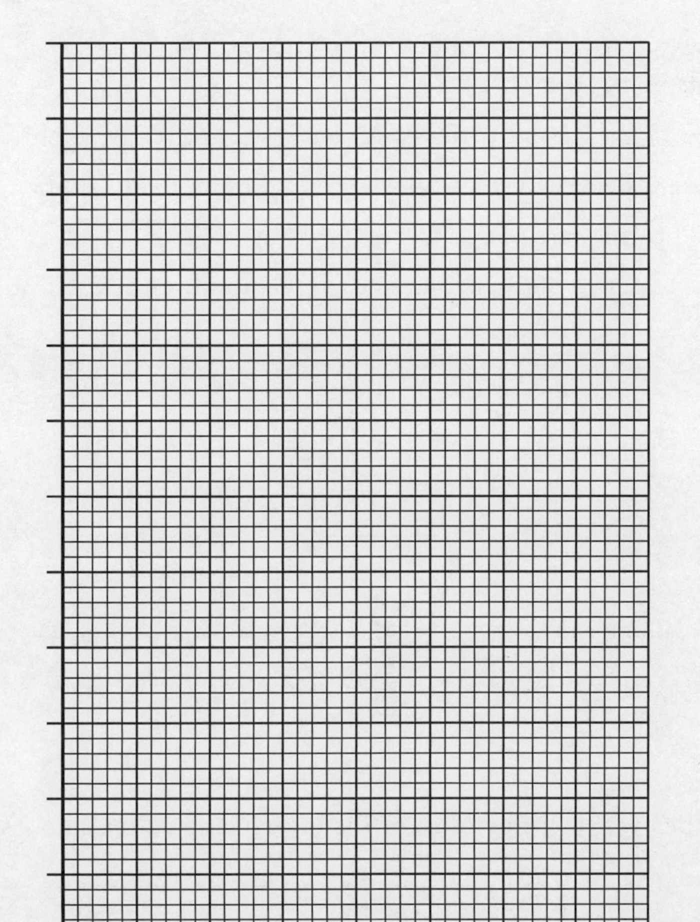

(*b*) (i) From the table, at which temperature was the number of yeast cells the highest after 2 hours?

_______ °C **1**

(ii) Suggest **one** improvement that would make this investigation **valid**.

____

____ **1**

[Turn over

DO NOT WRITE IN THIS MARGIN

Marks

9. (*a*) The black chemical on photographic film is stuck onto the film with a protein glue.

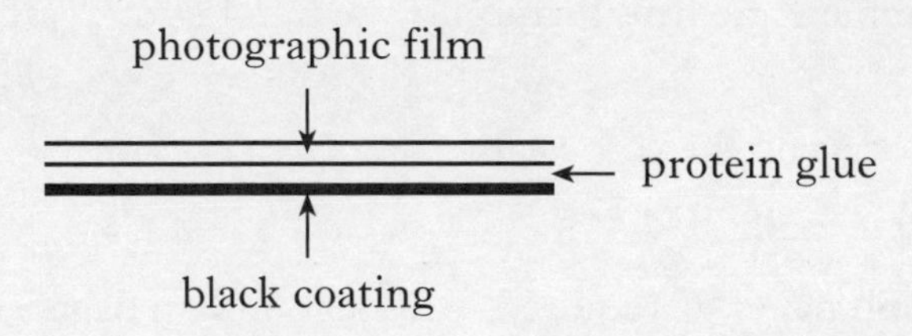

The following steps were carried out to investigate the effect of **different detergents** on the protein glue.

1 A piece of photographic film was placed in each of five identical test tubes.

2 A different detergent solution was poured into each test tube.

3 The test tubes were placed in a water bath at 40 °C.

4 The appearance of each piece of photographic film was noted after two hours.

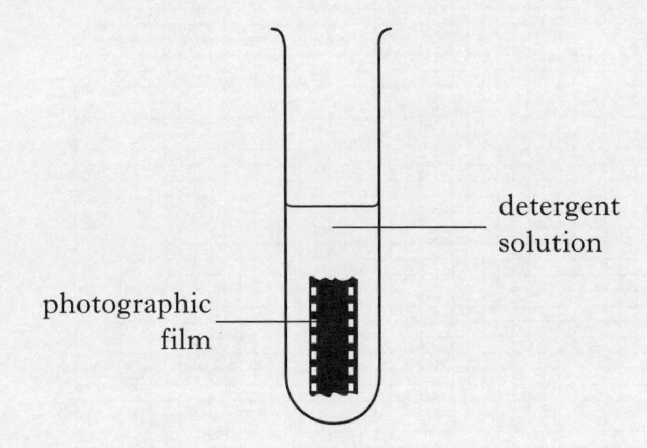

The table below shows the result of the investigation after 2 hours.

Detergent	Momo	Novo	Ohno	Promo	Quango
Appearance of photographic film	clear	black	black	clear	clear

(i) Identify the **biological** detergents used in the investigation.

____ **1**

(ii) Suggest an improvement which could be made to make the results of the investigation more **reliable**.

____

____ **1**

DO NOT WRITE IN THIS MARGIN

Marks

9. (continued)

(*b*) Some people are allergic to the enzymes in biological detergents.

How does the manufacturer reduce the chance of a detergent causing an allergic reaction?

____

____ **1**

(*c*) State **one** possible advantage of using a **biological** detergent.

____ **1**

[Turn over

DO NOT WRITE IN THIS MARGIN

Marks

10. (*a*) The diagram shows a method for the production of a fermented milk drink.

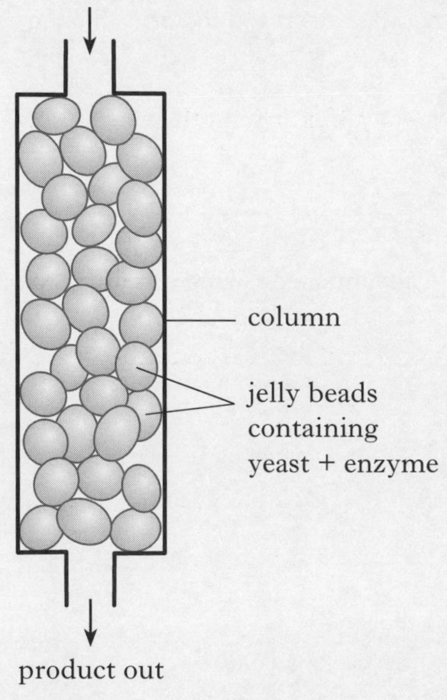

(i) Name the technique by which the yeast and enzyme are trapped in the jelly beads.

____ **1**

(ii) State **one** advantage of using this method of production.

____ **1**

(*b*) Fermentation also occurs during the brewing of beer.

Complete the summary of this process below.

Sugar ⟶ Alcohol + ____________________ **1**

DO NOT WRITE IN THIS MARGIN

Marks

10. (continued)

(*c*) **Underline** one option in each of the brackets to make the following sentences correct.

Cask conditioned beer differs from other beers because the yeast { is / is not } removed from the cask.

Brewery conditioned beer has { oxygen / carbon dioxide } added. **1**

(*d*) The disposal of waste produced in brewing can cause environmental problems.

State **one** way of avoiding such problems.

____

____ **1**

[Turn over

DO NOT WRITE IN THIS MARGIN

Marks

11. (*a*) The diagram below shows seedlings being pricked out and transferred to a tray.

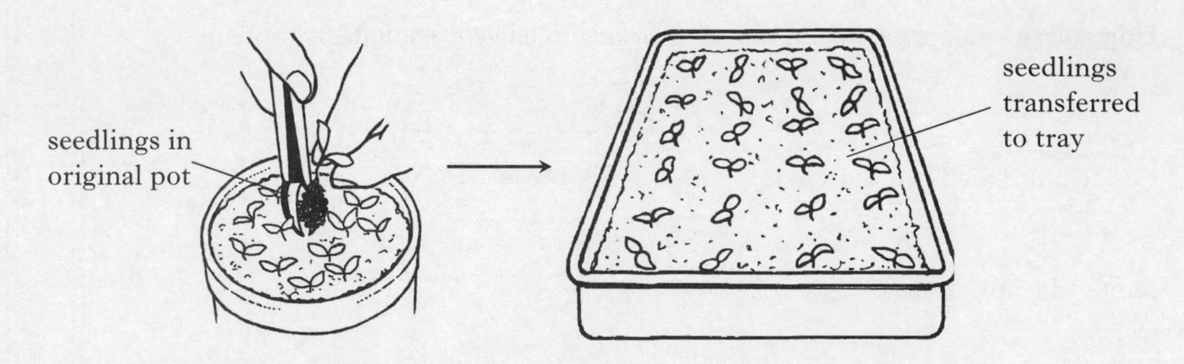

Give **one** reason why this procedure is carried out.

____

____ **1**

(*b*) New plants can be grown from plant propagation structures such as the food storage organ shown below.

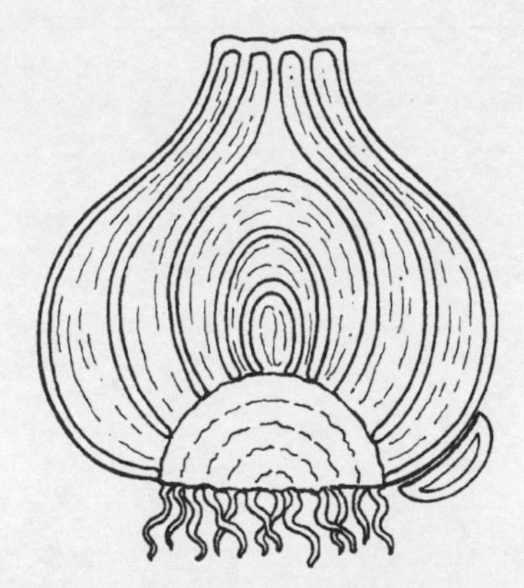

(i) Name this type of food storage organ.

____________________ **1**

(ii) Name **one other** plant propagation structure.

____________________ **1**

DO NOT WRITE IN THIS MARGIN

Marks

11. (continued)

(*c*) The following diagram shows a method of artificial propagation.

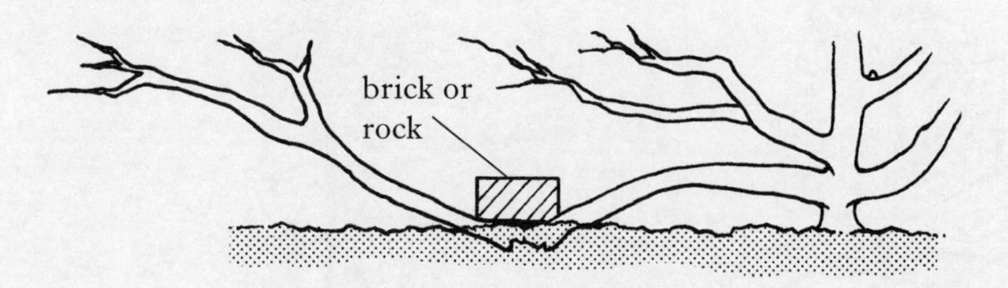

(i) Name this method of artificial propagation.

______________________________ **1**

(ii) State **one** advantage of this method of artificial propagation.

____

______________________________ **1**

[Turn over

DO NOT WRITE IN THIS MARGIN

Marks

12. The diagram shows a garden with a greenhouse and an automatic irrigation system.

(*a*) (i) What is supplied to the plants in this garden by the irrigation system?

______________________________ **1**

(ii) Suggest **one** reason for growing plants in the greenhouse.

______________________________ **1**

(*b*) Plants can be protected by floating fleece.

The following table shows the yields of various crops when grown with and without floating fleece.

Crop	*Average yield* (units per hectare)	
	Without *floating fleece*	***With*** *floating fleece*
cantaloupe	1500	6000
cucumber	240	720
pepper	300	900
tomato	800	2400
watermelon	750	2500

(i) Which crop shows the greatest increase in yield (units per hectare) when grown with floating fleece?

Space for calculation

______________________ **1**

DO NOT WRITE IN THIS MARGIN

Marks

12. (*b*) (continued)

(ii) On the grid below complete the bar graph by

(1) putting a scale on the vertical axis **1**

(2) plotting the remaining results for the average yield with floating fleece. **1**

(Additional graph paper, if required, will be found on page 34.)

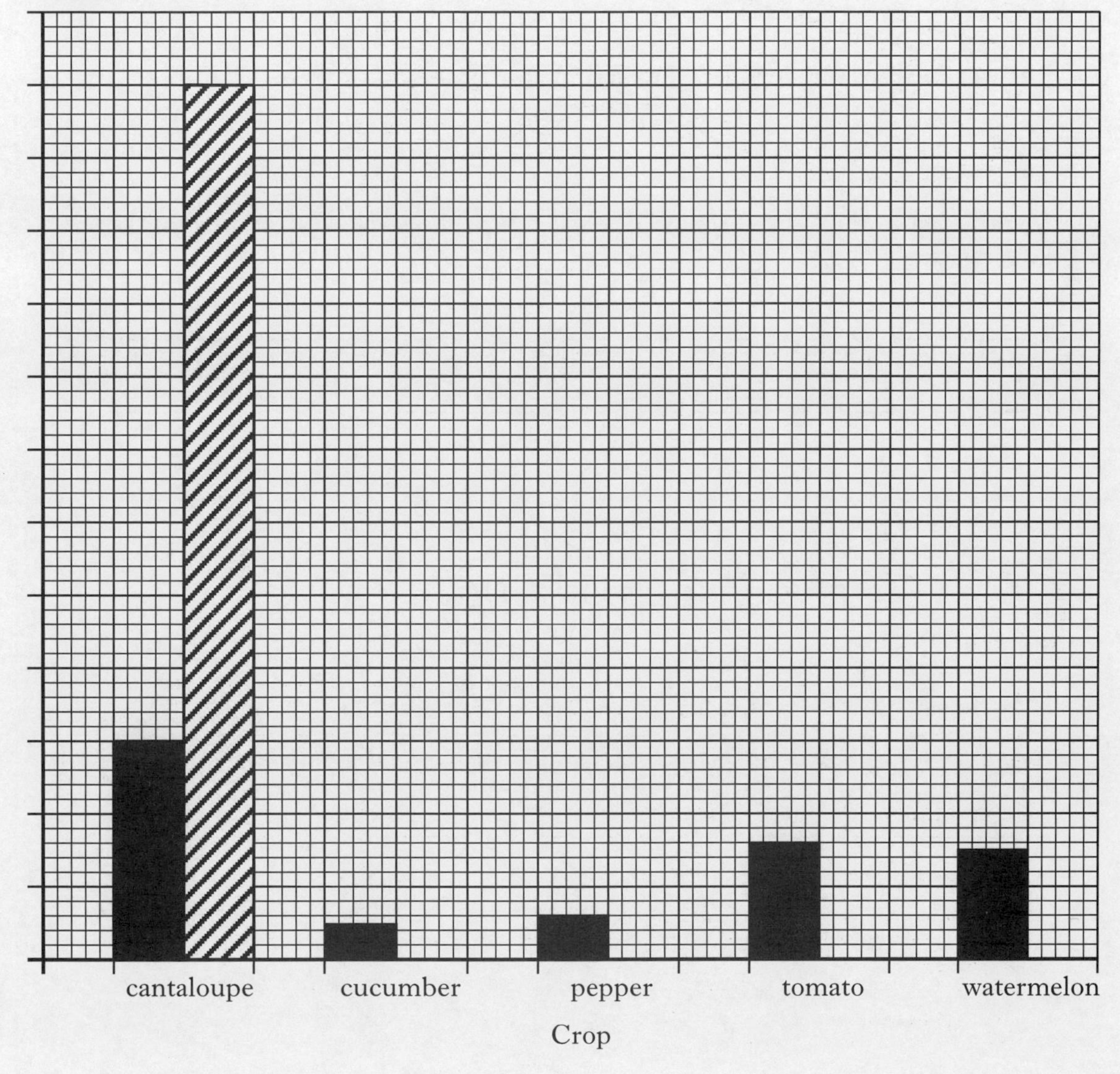

[Turn over

DO NOT WRITE IN THIS MARGIN

Marks

13. A student carried out an investigation to compare the germination of four different types of seeds.
The results are shown in the table below.

Type of seed	*Number of seeds sown*	*Number of seeds germinated*
sunflower	28	15
grass	20	18
lettuce	24	12
tomato	20	12

(*a*) Half the seeds of one type failed to germinate. Name this type of seed.

______________________ **1**

(*b*) Identify **two** variables that should have been kept the same when setting up the investigation.

1 ____

2 ____ **2**

[*END OF QUESTION PAPER*]

DO NOT WRITE IN THIS MARGIN

SPACE FOR ANSWERS

ADDITIONAL PIE CHART FOR QUESTION 4(*a*)

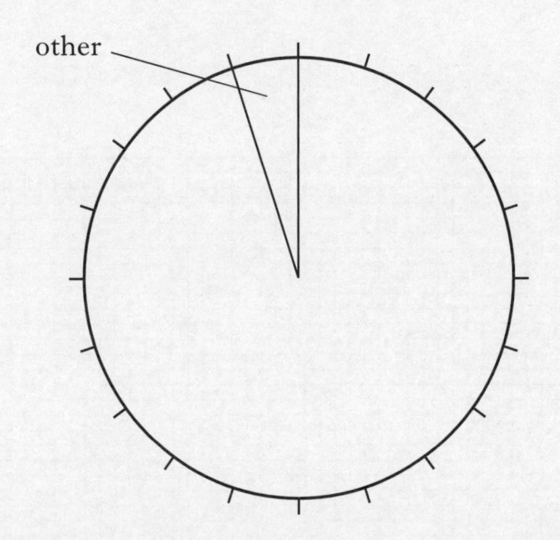

ADDITIONAL GRAPH PAPER FOR QUESTION 8(*a*)

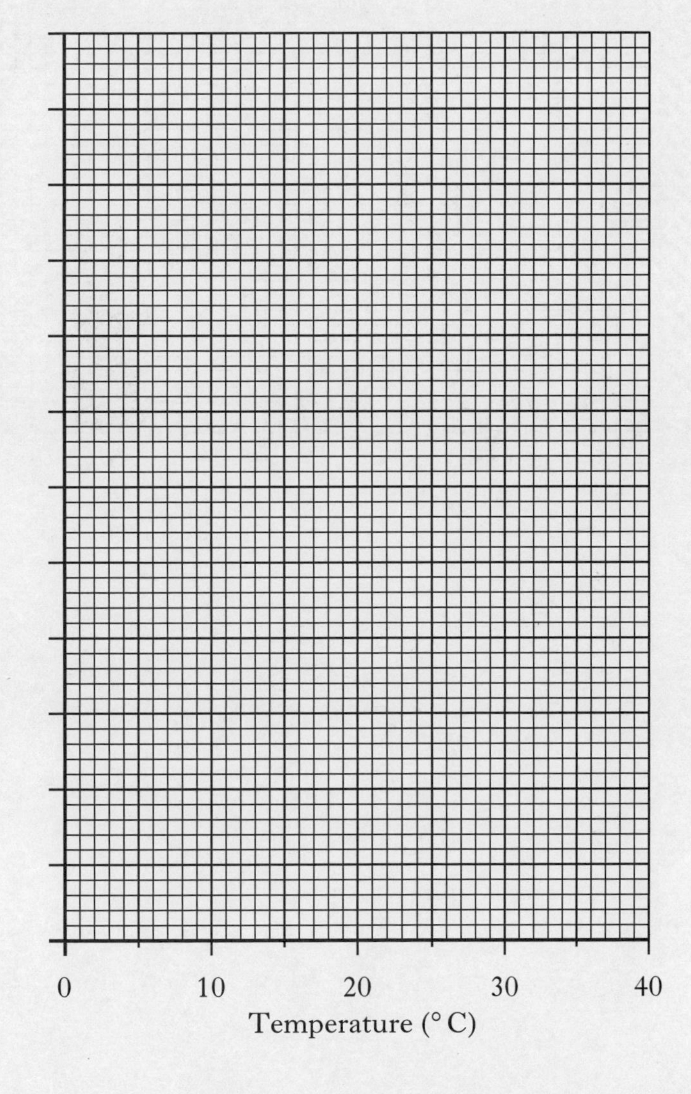

DO NOT WRITE IN THIS MARGIN

SPACE FOR ANSWERS

ADDITIONAL GRAPH PAPER FOR QUESTION 12(*b*)(ii)

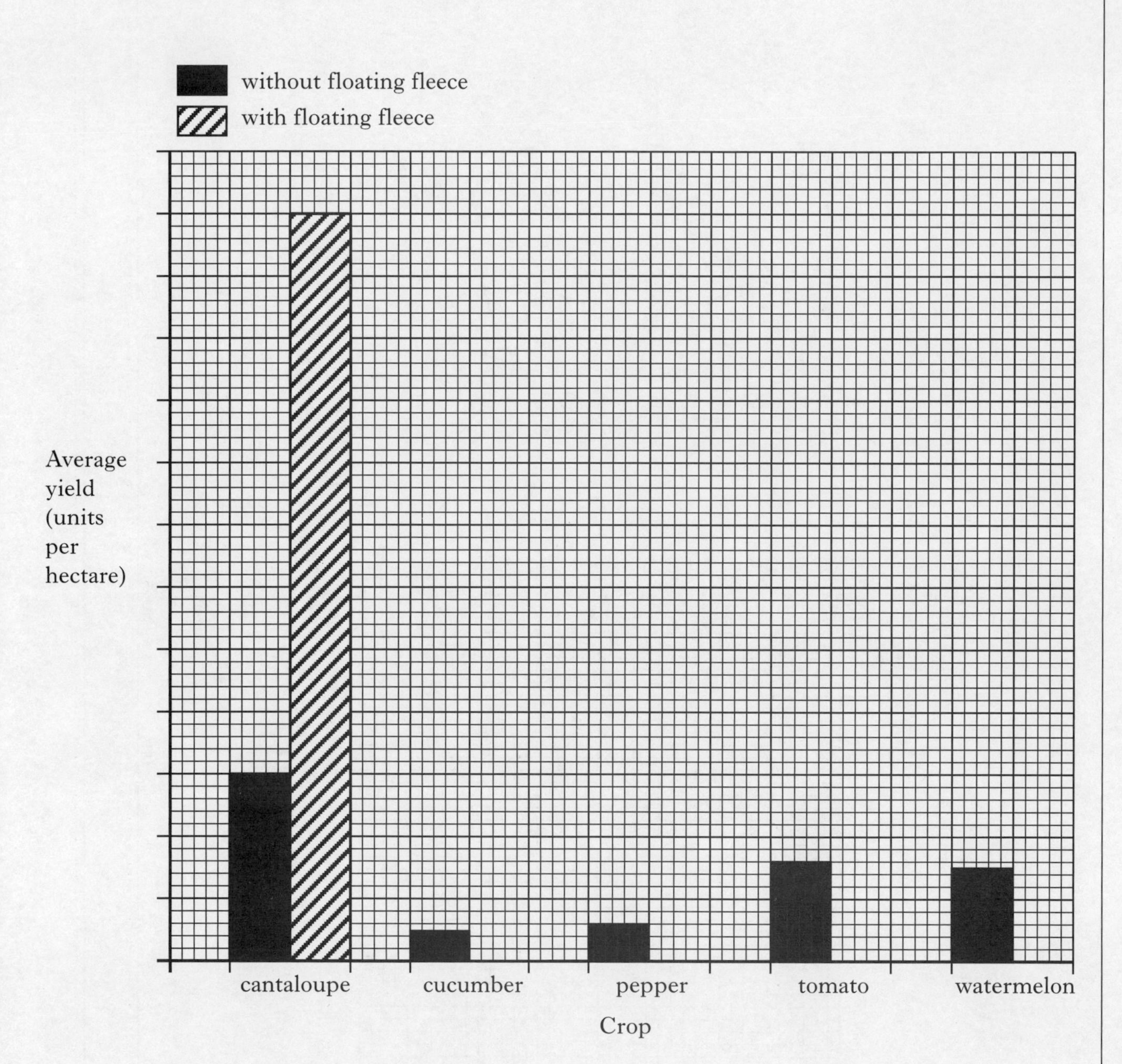

DO NOT WRITE IN THIS MARGIN

SPACE FOR ANSWERS

DO NOT WRITE IN THIS MARGIN

SPACE FOR ANSWERS

[BLANK PAGE]

FOR OFFICIAL USE

Section B Total

X007/101

NATIONAL QUALIFICATIONS 2006

TUESDAY, 23 MAY
9.00 AM – 10.30 AM

BIOLOGY
INTERMEDIATE 1

Fill in these boxes and read what is printed below.

Full name of centre

Town

Forename(s)

Surname

Date of birth
Day Month Year

Scottish candidate number

Number of seat

SECTION A

Instructions for completion of Section A are given on page two.

For this section of the examination you must use an **HB pencil**.

SECTION B

1 All questions should be attempted.

2 The questions may be answered in any order but all answers are to be written in the spaces provided in this answer book, **and must be written clearly and legibly in ink**.

3 Additional space for answers will be found at the end of the book. If further space is required, supplementary sheets may be obtained from the invigilator and should be inserted inside the **front** cover of this book.

4 The numbers of questions must be clearly inserted with any answers written in the additional space.

5 Rough work, if any should be necessary, should be written in this book and then scored through when the fair copy has been written. If further space is required, a supplementary sheet for rough work may be obtained from the invigilator.

6 Before leaving the examination room you must give this book to the invigilator. If you do not, you may lose all the marks for this paper.

LI X007/101 6/7320

Read carefully

1 Check that the answer sheet provided is for **Biology Intermediate 1 (Section A)**.

2 For this section of the examination you must use an **HB pencil** and, where necessary, an eraser.

3 Check that the answer sheet you have been given has **your name**, **date of birth**, **SCN** (Scottish Candidate Number) and **Centre Name** printed on it.

Do not change any of these details.

4 If any of this information is wrong, tell the Invigilator immediately.

5 If this information is correct, **print** your name and seat number in the boxes provided.

6 The answer to each question is **either** A, B, C or D. Decide what your answer is, then, using your pencil, put a horizontal line in the space provided (see sample question below).

7 There is **only one correct** answer to each question.

8 Any rough working should be done on the question paper or the rough working sheet, **not** on your answer sheet.

9 At the end of the exam, put the **answer sheet for Section A inside the front cover of this answer book**.

Sample Question

Which of the following foods contains a high proportion of food?

A Butter

B Bread

C Sugar

D Apple

The correct answer is **A**—Butter. The answer **A** has been clearly marked in **pencil** with a horizontal line (see below).

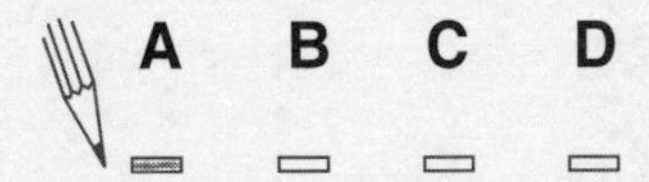

Changing an answer

If you decide to change your answer, carefully erase your first answer and using your pencil, fill in the answer you want. The answer below has been changed to **D**.

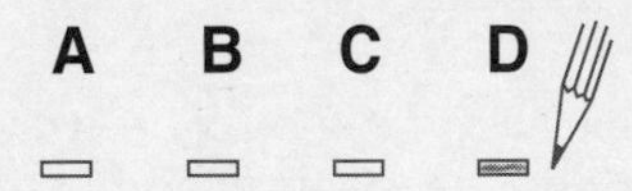

SECTION A

All questions in this Section should be attempted.

Answers should be given on the separate answer sheet provided.

1. The following diagram shows the structure of a broad bean seed.

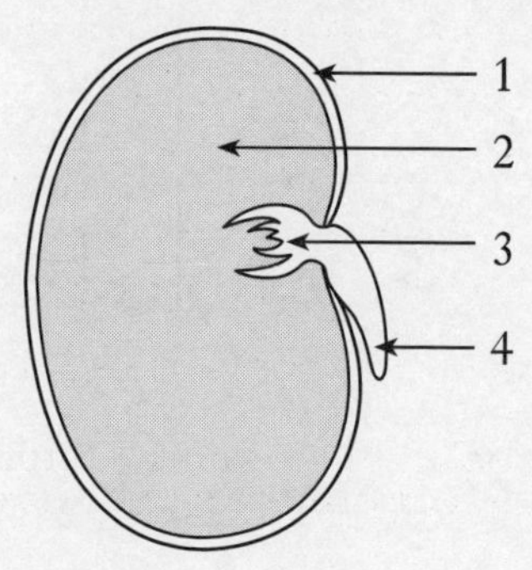

The embryo plant is shown by label(s)

A 1 only

B 1 and 2

C 3 and 4

D 4 only.

2. Which of the following is **not** required for the germination of seeds?

A Water

B A suitable temperature

C Oxygen

D Fertiliser

3. The information from a seed packet is shown below.

Jan Feb Mar Apr May Jun Jul Aug Sep Oct Nov Dec

Key

Sow

Plant out

Flower

Which of the following is correct for this type of seed?

	Sow	*Plant out*	*Flower*
A	February/March	April	September/October
B	February/March	May	September/October
C	May	February/March	September/October
D	September/October	May	February/March

4. A group of students set up the following investigation.

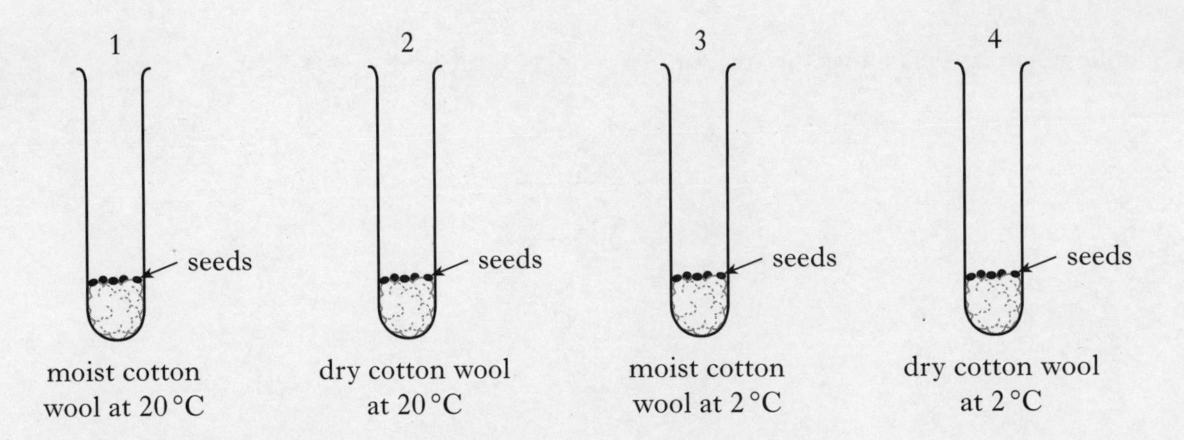

Which two tubes would be used to investigate the effect of temperature on germination?

A 1 and 2

B 2 and 3

C 1 and 3

D 3 and 4

5. Which of the following is used to protect plants as they grow?

A Capillary matting

B Water retentive gel

C Floating fleece

D Automatic irrigation

6. Several animal pests may be found in a greenhouse.

Use the following key to identify the pest shown.

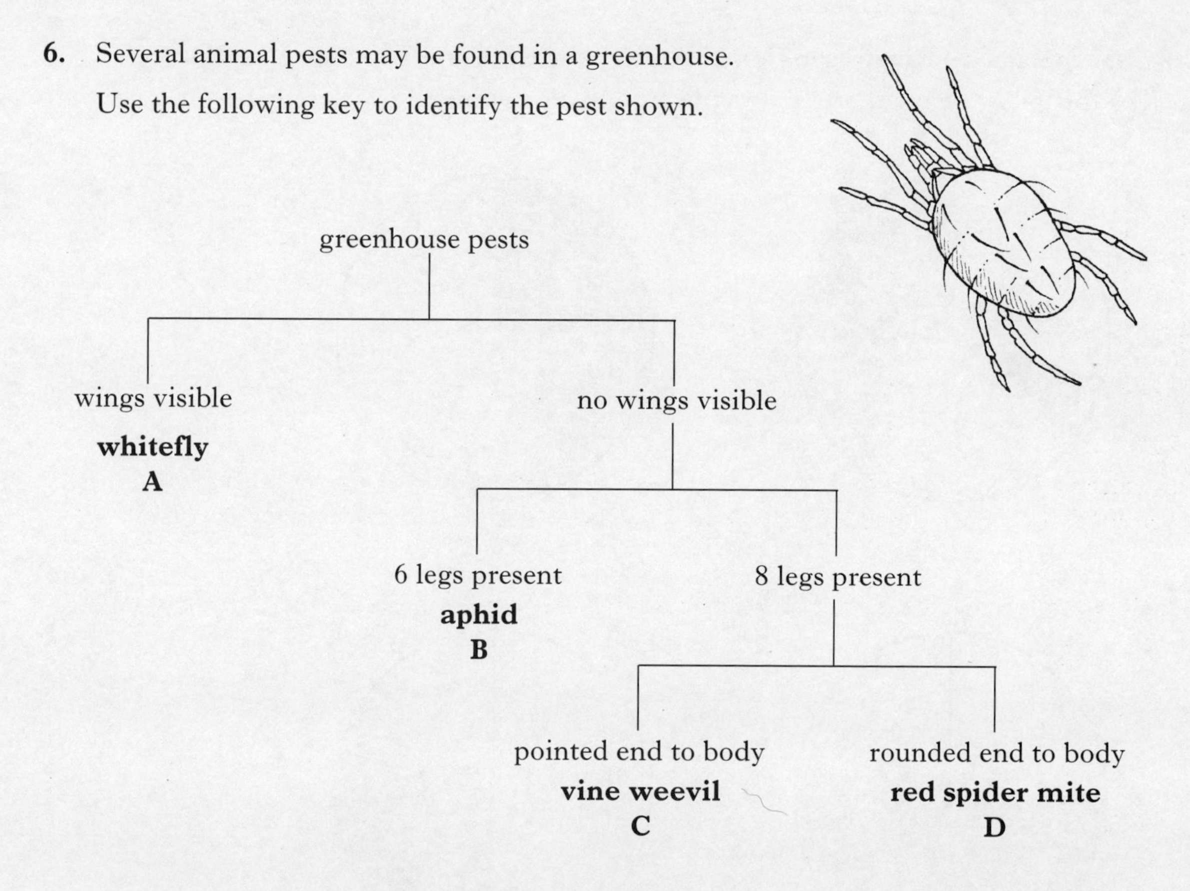

7. A soil test kit can be used to determine soil type.

The colour range of the kit is shown in the table.

Colour	*Soil type*
orange	strongly acidic
yellow	weakly acidic
bright green	neutral
dark green	weakly alkaline
blue	strongly alkaline

A gardener tested four samples of soil from four different sites.

The results of the soil tests are shown below.

Soil site	*Colour*
A	yellow
B	blue
C	orange
D	dark green

Heather plants grow best in weakly acidic soils.

Which soil is the best for growing heathers?

8. A fertiliser contained 40% nitrogen (N), 25% phosphorus (P) and 5% potassium (K).
Which of the following pie charts shows this information correctly?

A

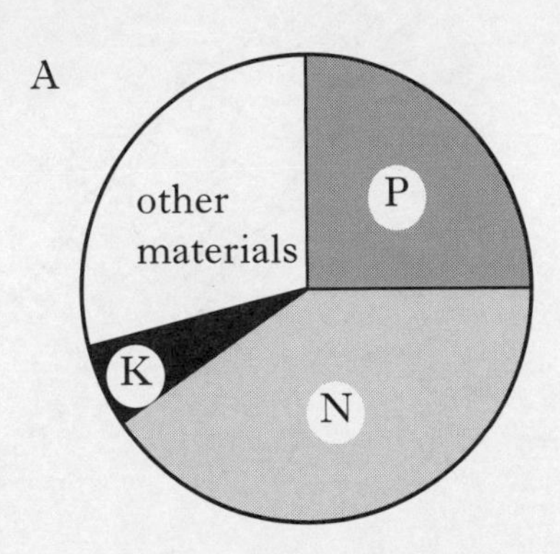

B

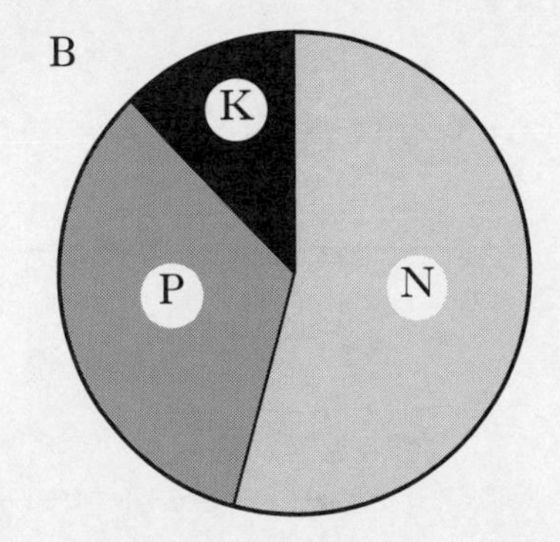

C

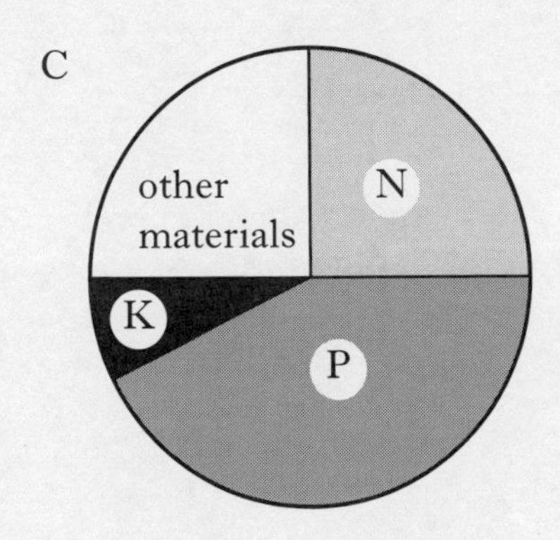

D

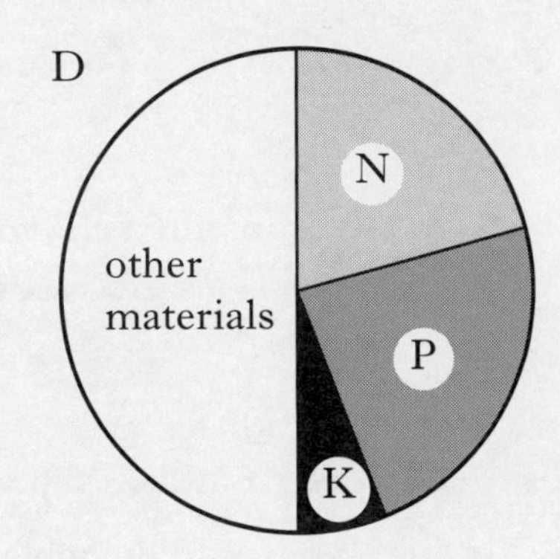

9. Washing powders described as "biological" contain

A bacteria

B fungi

C enzymes

D antibiotics.

10. An experiment was set up to compare stain removal by two biological detergents.

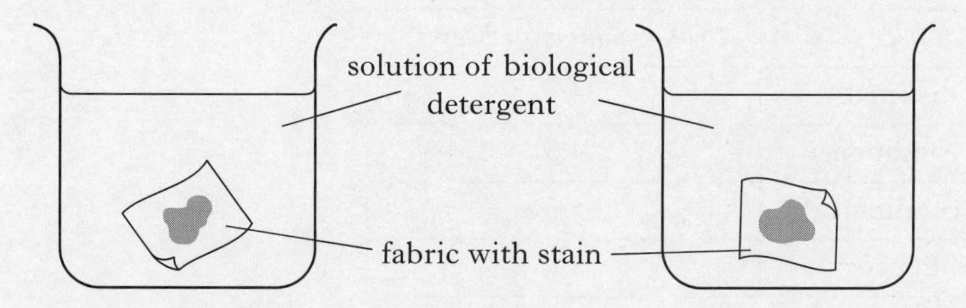

Which of the following factors should be kept constant?

R – type of biological detergent
S – type of fabric
T – type of stain

A S and T only

B R and T only

C R and S only

D R, S and T

11. Why have some bacteria become resistant to antibiotics?

A The range of antibiotics has increased.

B Antibiotics have been overused.

C The antibiotics are produced by fungi.

D Too few antibiotics have been prescribed by doctors.

[Turn over

12. The alcohol content of four types of beer are shown in the following table.

Type of beer	*Alcohol content of beer* (%)
W – cask conditioned	4·4
X – brewery conditioned	4·3
Y – brewery conditioned	5·5
Z – cask conditioned	3·8

Which line in the table below correctly shows the average alcohol content of the brewery conditioned and cask conditioned beers?

	Average alcohol content (%)	
	Brewery conditioned beers	*Cask conditioned beers*
A	4·9	4·1
B	4·1	4·9
C	9·8	8·2
D	8·2	9·8

13. Which of the following is removed from brewery conditioned beer?

A Carbon dioxide

B Yeast

C Alcohol

D Oxygen

14. The resazurin test is carried out on milk to detect

A bacteria

B viruses

C sugar

D protein.

15. Salmon in fish farms are fed with yeast products to produce flesh with

A less fat

B more protein

C better flavour

D pink colour.

16. The diagram below shows an investigation into the production of a fermented drink.

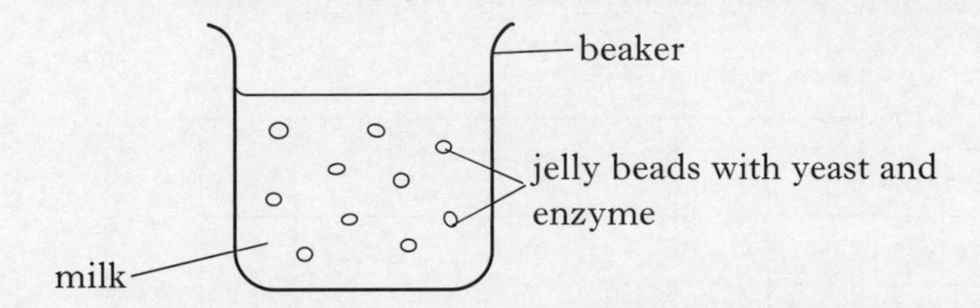

A student predicts that yeast and an enzyme are needed for the milk to be fermented.

Which of the following would be a suitable control for this investigation?

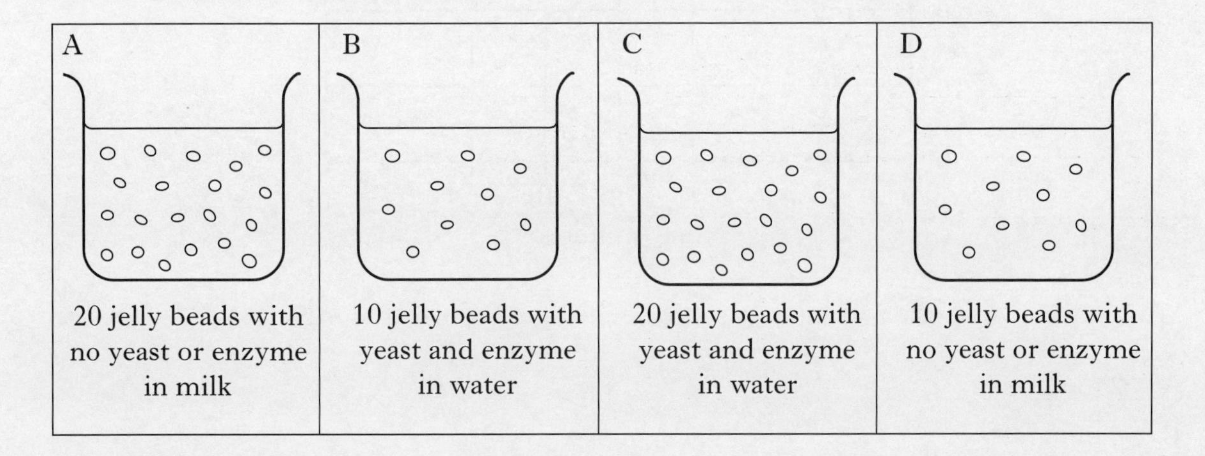

17. What name is given to the technique in which yeast and enzymes are trapped in jelly beads?

A Fermentation

B Immobilisation

C Pasteurisation

D Separation

[Turn over

18. The pulse rate of a fifteen-year-old girl was recorded before, during and after a two minute period of exercise. The results are presented in the graph below.

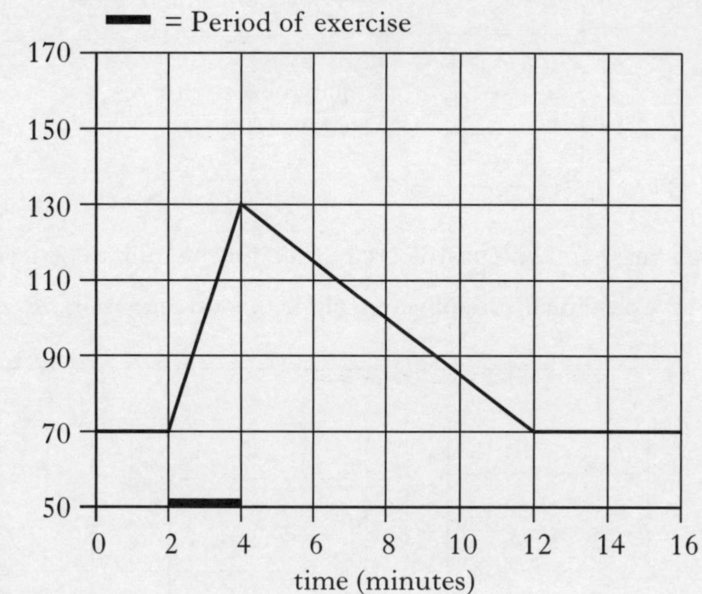

What was the time taken for her pulse rate to return to normal after exercise had stopped?

A 2 minutes

B 8 minutes

C 10 minutes

D 12 minutes

19. Normal body temperature is

A 30 °C

B 35 °C

C 37 °C

D 40 °C.

20. Smoking tobacco increases the risk of

A hypothermia

B heart disease

C arthritis

D anaemia.

21. In an investigation to measure reaction time, a student was asked to press a buzzer as soon as he saw the light going on as shown in the diagram below.

The results are shown in the table.

Reaction time (milliseconds)	
After 1 trial	*After 10 trials*
500	300

After 10 trials, the percentage decrease in the student's reaction time was

A 30%

B 40%

C 50%

D 60%.

22. Which physiological measurement could be used to detect leukaemia?

A Body fat

B Blood pressure

C Blood cell count

D Pulse rate

[Turn over

23. Which of the following sequences correctly represents the pathway taken by air when breathing in?

A Bronchus ⟶ windpipe ⟶ air sac ⟶ bronchiole

B Windpipe ⟶ bronchus ⟶ air sac ⟶ bronchiole

C Bronchus ⟶ windpipe ⟶ bronchiole ⟶ air sac

D Windpipe ⟶ bronchus ⟶ bronchiole ⟶ air sac

24. Which two food groups provide energy?

A Vitamins and minerals

B Minerals and fats

C Proteins and vitamins

D Carbohydrates and fats

25. A student tested four foods, A, B, C and D, for the presence of starch, sugar and protein.

The tests used were:

Starch present – iodine solution turns from brown to black.

Sugar present – clinistix turns from pink to purple.

Protein present – albustix turns from yellow to green.

The results are shown in the table below.

	Colour produced		
Food	*Starch test*	*Sugar test*	*Protein test*
A	black	purple	green
B	black	pink	yellow
C	brown	purple	yellow
D	brown	purple	green

Which food contained starch, sugar and protein?

Candidates are reminded that the answer sheet for Section A MUST be returned inside this answer book.

[Turn over for Section B on *Page fourteen*

DO NOT WRITE IN THIS MARGIN

Marks

SECTION B

All questions in this Section should be attempted.

All answers must be written clearly and legibly in ink.

1. (*a*) **Read the following passage carefully.**

HOW TO AVOID DAMPING OFF

Adapted from *The Greenock Telegraph*

Damping off is one of the most common plant diseases. It not only causes problems on flower seedlings such as *Lobelia, Petunia* and *Salvia* but also affects vegetable plants including cabbage, cauliflower and lettuce.

The symptoms of damping off disease can be varied. In some cases the seedlings become weak looking and shrivel up or simply collapse and die as shown in the picture below. In others the root system will rot away.

To avoid damping off it is necessary to clean and sterilise all trays and pots as well as greenhouse benches. Old compost should not be used for sowing and potting on. Good clean tap water, to which a fungicide may be added, must be used for watering the seeds after sowing and pricking out.

It is also important that seedlings are pricked out as soon as possible after germination. However, if the disease affects seedlings, they should be disposed of together with the compost in which they were growing.

Answer the questions below, using information from the passage.

(i) Name **one** flower and **one** vegetable affected by damping off disease.

Flower ______________________________

Vegetable ______________________________ **1**

(ii) Give **two** symptoms of the disease.

1 ______________________________

2 ______________________________ **2**

DO NOT WRITE IN THIS MARGIN

Marks

1. (*a*) (continued)

(iii) Describe **one** method of preventing damping off disease.

____

____ **1**

(iv) What should be done to seedlings as soon as possible after they have germinated?

____ **1**

(*b*) The composition of a compost is shown below.

	Percentage
Loam	75
Peat	20
Sand	5

(i) Present this information as a pie chart.

(An additional pie chart, if needed, will be found on page 28.)

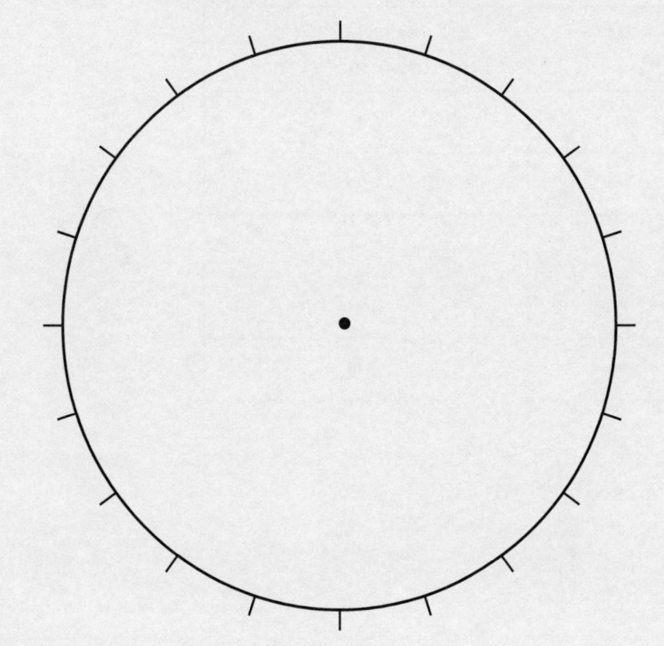

2

(ii) What can be added to compost to improve drainage?

____ **1**

DO NOT WRITE IN THIS MARGIN

Marks

2. (*a*) A student carried out an investigation into the effect of water on seed germination.

Five dishes were set up as shown below.

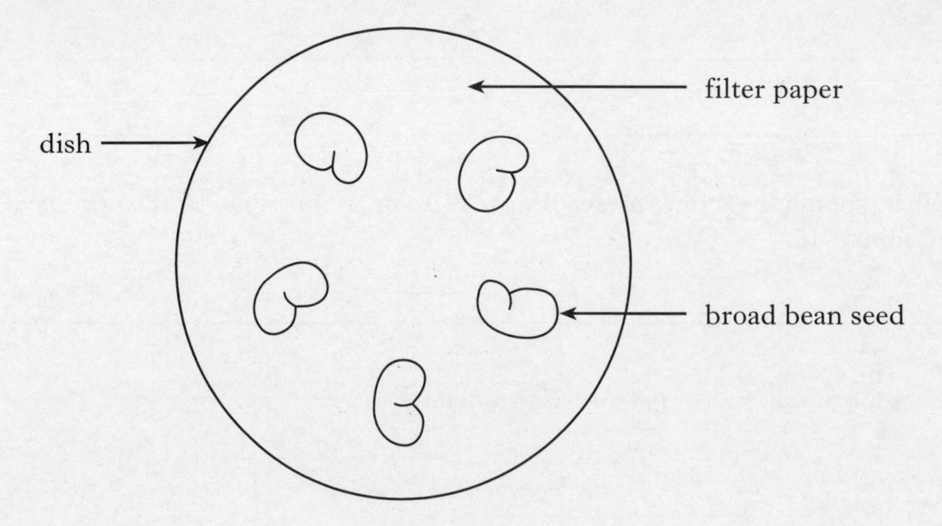

A different volume of water was added to each dish.

The dishes were covered and placed in a cupboard at room temperature.

After ten days, the length of the seedling roots were measured and the average length calculated.

The results are shown below.

Volume of water added (cm^3)	*Average length of seedling root* (cm)
0	0
5	1·6
10	3·4
15	6·6
20	2·0

(i) Using the information in the table, state the volume of water which resulted in the greatest root growth.

________________ cm^3 **1**

DO NOT WRITE IN THIS MARGIN

Marks

2. (*a*) (continued)

(ii) On the grid below, complete the **line graph** by

(1) providing a label on the horizontal axis **1**

(2) providing a scale on the vertical axis **1**

(3) plotting the results. **1**

(Additional graph paper, if required, will be found on page 28.)

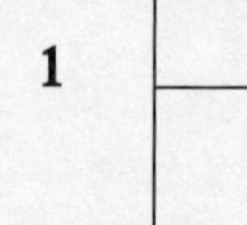

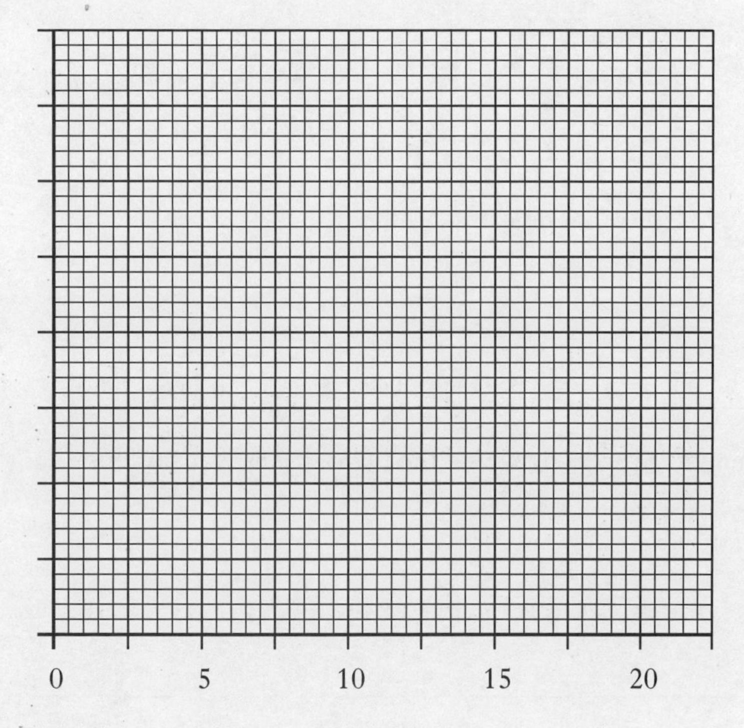

(iii) Five seeds were used in each dish.

Explain why this was good experimental technique.

____

____ **1**

(*b*) Broad bean seeds are large and easy to sow individually.

Describe **one** way in which very small seeds can be sown.

____

____ **1**

[Turn over

DO NOT WRITE IN THIS MARGIN

Marks

3. (*a*) New potato plants can be grown from the storage organs shown below.

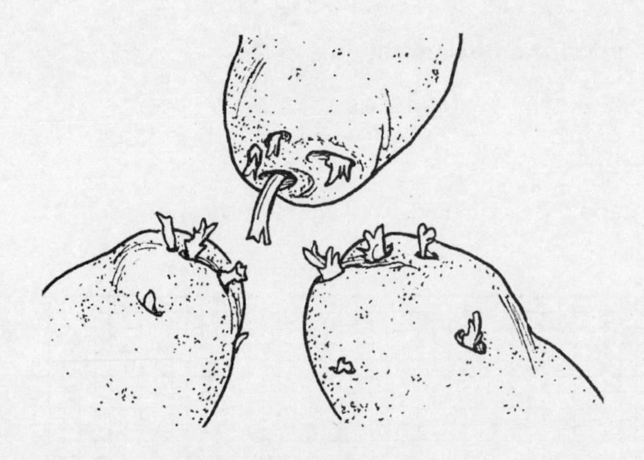

Name this type of storage organ.

________________________________ **1**

(*b*) Once the potato plant has developed leaves, it can make its own food.

(i) Name the process by which the plant makes its own food.

________________________________ **1**

(ii) State **one** use the plant makes of the food produced in this way.

____ **1**

(*c*) The diagram below shows another method of plant propagation.
Use the words from the list to label the diagram.

List: runner plantlet parent plant

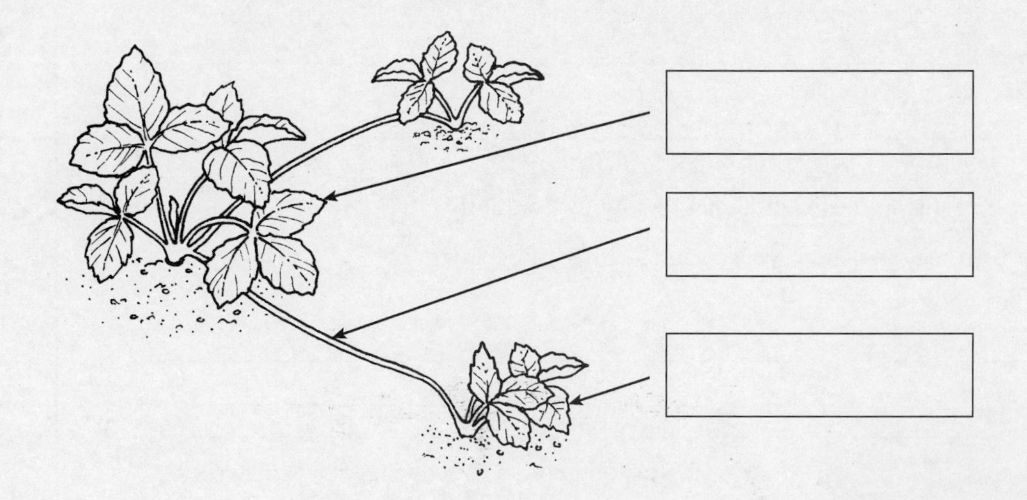

1

DO NOT WRITE IN THIS MARGIN

Marks

4. The diagrams below show a method of growing new plants without using seeds.

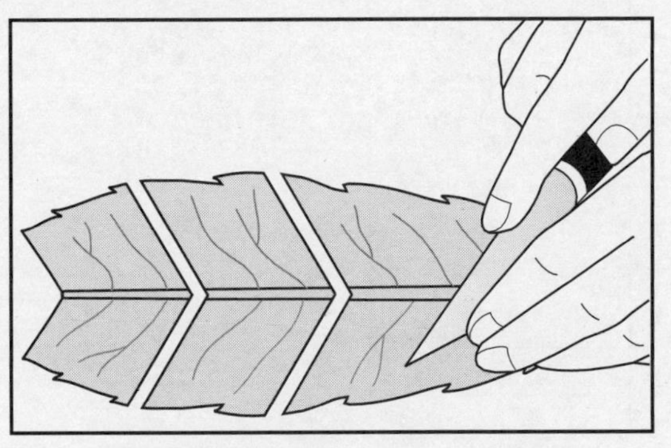

1. Cut V-shaped parts out of leaf

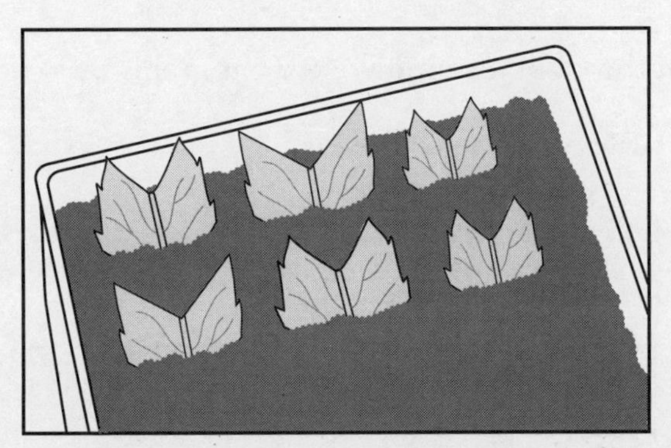

2. Place leaf parts into soil in tray

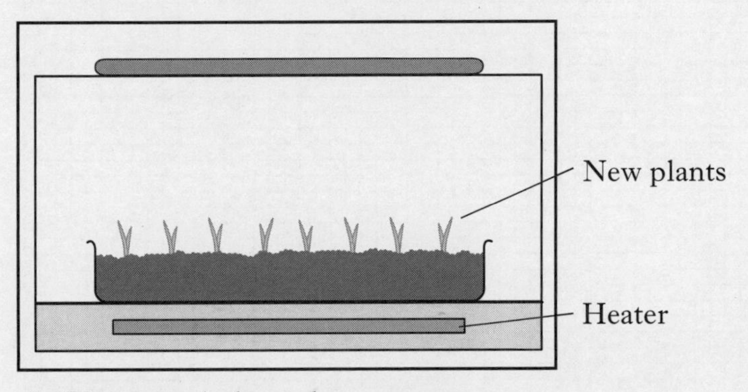

3. Place tray in heated propagator

(*a*) What name is given to this method of growing new plants?

______________________________ **1**

(*b*) State **one** advantage of using the heated propagator.

____

______________________________ **1**

(*c*) Describe **one** additional step that could be taken to encourage root growth.

____

______________________________ **1**

DO NOT WRITE IN THIS MARGIN

Marks

5. The label below is from a carton of milk.

Content	*Mass per 100g* (g)
Water	87
Protein	4
Fat	4
Sugar	5

(*a*) Use the information from the label to complete the bar graph by

(i) putting a label and scale on the vertical axis **1**

(ii) plotting the other bars. **1**

(Additional graph paper, if required, will be found on page 29.)

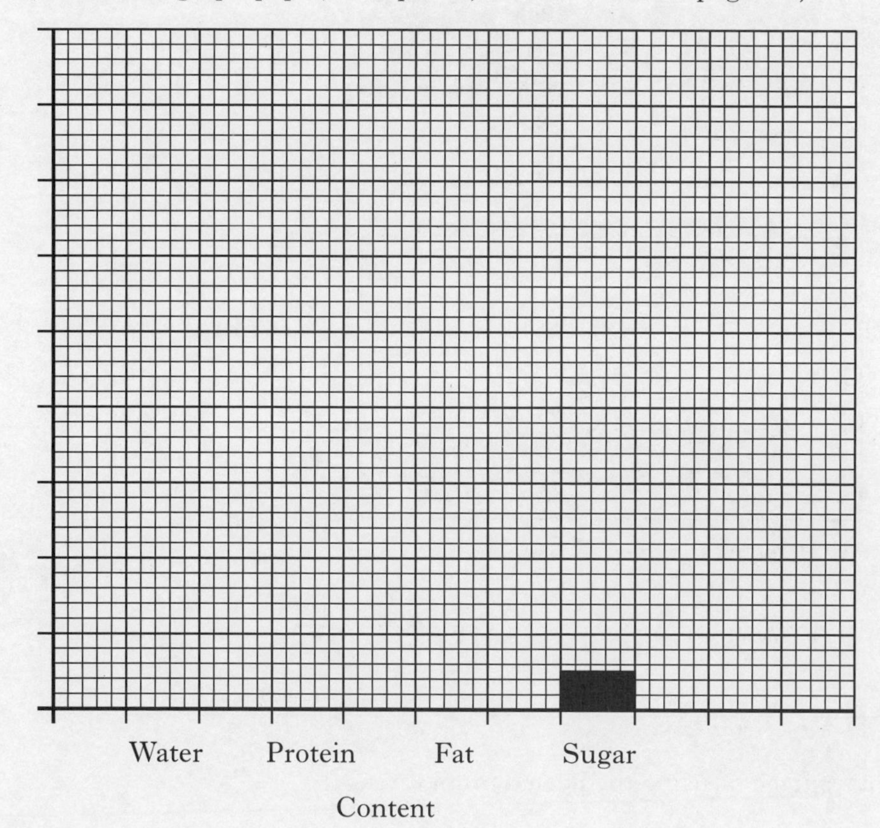

(*b*) Why is milk pasteurised?

______________________________ **1**

(*c*) Name **one** method of preserving fresh milk.

______________________________ **1**

DO NOT WRITE IN THIS MARGIN

Marks

5. (continued)

(*d*) The Recommended Daily Allowances (RDA) of fat, sugar and fibre for 15–17 year olds are shown in the table.

	Fat	*Sugar*	*Fibre*
RDA (g)	25	5	30

(i) A 100 g carton of yoghurt contains 2·5 g of fat.

What percentage of the RDA of fat would be provided by a 100 g carton of yoghurt?

Space for calculation

__________ % **1**

(ii) Calculate the ratio of the RDA of sugar to the RDA of fibre.

Express your answer as a simple whole number.

Space for calculation

__________ : __________ **1**

sugar fibre

[Turn over

DO NOT WRITE IN THIS MARGIN

Marks

6. (*a*) A student carried out an investigation on cheese-making.

The time taken for rennet to clot milk protein at different temperatures is shown in the table.

Temperature (°C)	*Time taken for milk protein to clot* (min)
10	No clotting
20	35
30	8
40	3
50	50
60	No clotting

(i) At what temperature was clotting fastest?

__________ °C **1**

(ii) State a temperature that would **not** be suitable for this process.

__________ °C **1**

(*b*) Name the solid formed when milk protein clots.

______________________________ **1**

(*c*) (i) Tick (✓) the correct box in **each column** to show the effect of the disposal of waste whey into rivers.

Number of bacteria		*Oxygen level*		*Number of other organisms*	
Increases		Increases		Increases	
Decreases		Decreases		Decreases	

2

(ii) Describe **one** way in which whey can be upgraded.

____

______________________________ **1**

DO NOT WRITE IN THIS MARGIN

Marks

7. The graph shows the results of an investigation on the raising of bread dough.

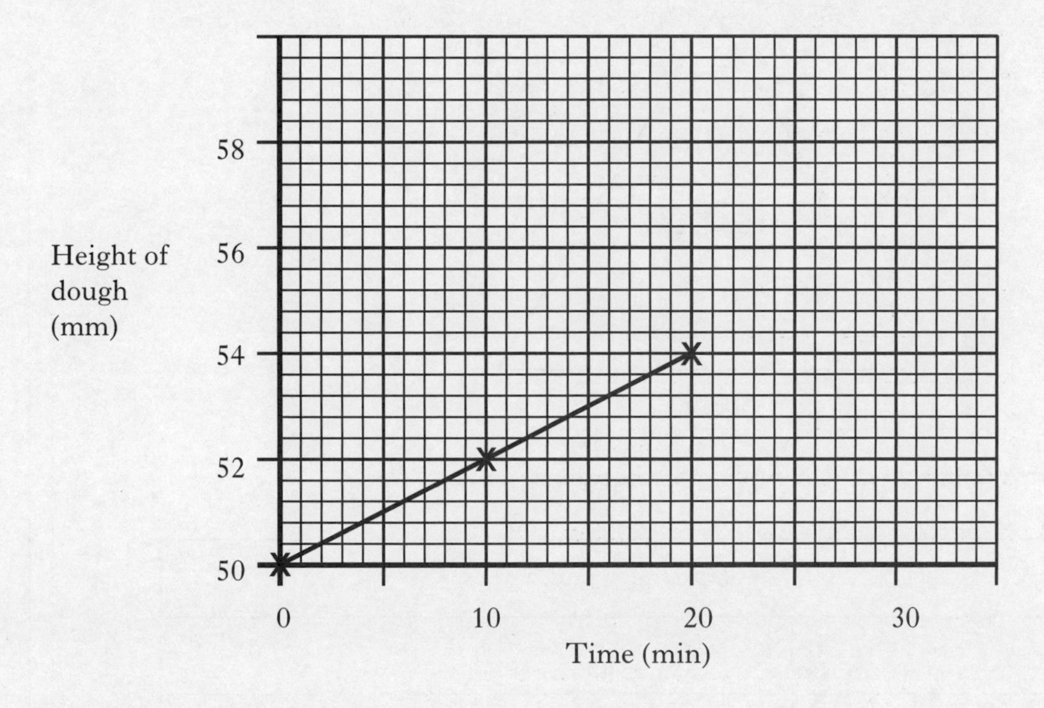

(*a*) **Use the information in the graph** to complete the table below.

Time (min)	0	10	
Height of dough (mm)			54

1

(*b*) Predict the height of the dough after **30 minutes**.

__________ mm **1**

(*c*) Name the gas produced which makes the dough rise.

____________________________________ **1**

[Turn over

DO NOT WRITE IN THIS MARGIN

Marks

8. (*a*) The three aspects of good health are shown below.

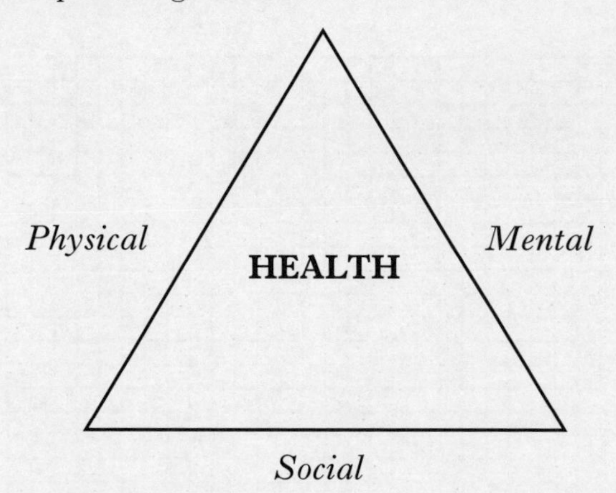

Use this information to complete the headings in the table.

Aspects of Health		
Exercise Diet	Feelings Emotions	Family Friends

1

(*b*) State **one** unnecessary health risk that should be avoided for a healthy lifestyle.

______________________________ **1**

DO NOT WRITE IN THIS MARGIN

Marks

9. The following graph shows the volume of air in the lungs of a female student over a period of thirty seconds.

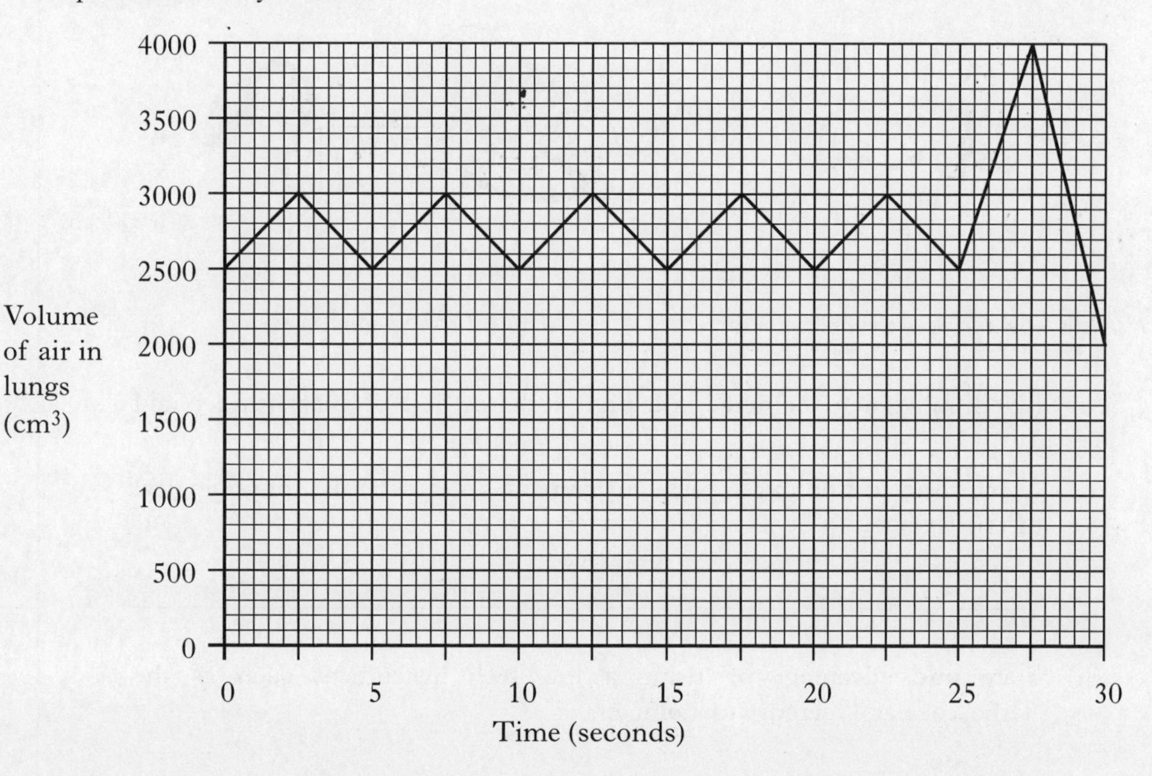

(*a*) Calculate the student's normal breathing rate in breaths **per minute**.

Space for calculation

____________ breaths per minute **1**

(*b*) Another physiological measurement was taken at 25 seconds.

The student was asked to breathe in as deeply as she could then breathe out as much air as possible in one breath.

What term is used to describe this measurement?

____ **1**

[Turn over

DO NOT WRITE IN THIS MARGIN

Marks

10. (*a*) The diagrams below show two methods of measuring blood pressure.

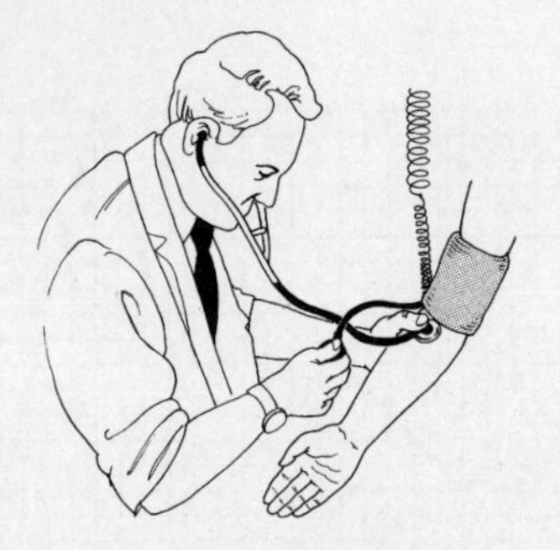

Stethoscope and mercury manometer

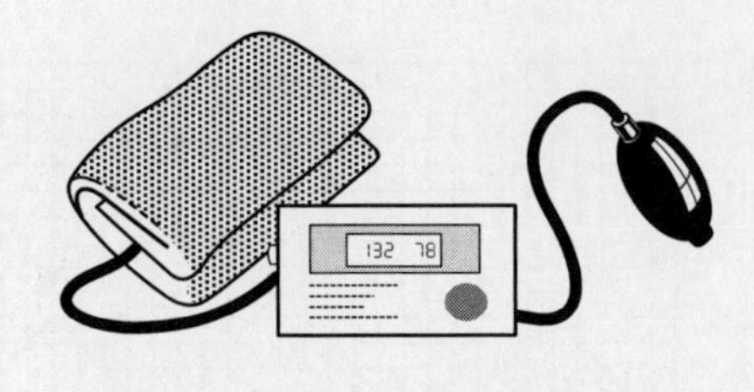

Digital sphygmomanometer

The digital sphygmomanometer is a high tech method of measuring blood pressure.

(i) Name **one other** high tech instrument used to take physiological measurements.

______________________________ 1

(ii) State **one** advantage of using a low tech instrument such as the stethoscope and mercury manometer.

______________________________ 1

(*b*) The table below refers to medical conditions that can be caused by high or low blood pressure.

Tick (✓) the correct box for each medical condition to show if it can be caused by high blood pressure or low blood pressure.

	Cause	
Medical condition	*High blood pressure*	*Low blood pressure*
Angina		
Stroke		
Fainting		
Heart attack		

2

(*c*) State **one** factor which can affect pulse rate.

______________________________ 1

(*d*) Name **one** substance carried by the blood.

______________________________ 1

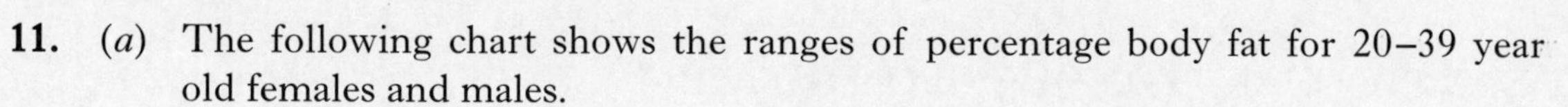

DO NOT WRITE IN THIS MARGIN

Marks

11. (*a*) The following chart shows the ranges of percentage body fat for 20–39 year old females and males.

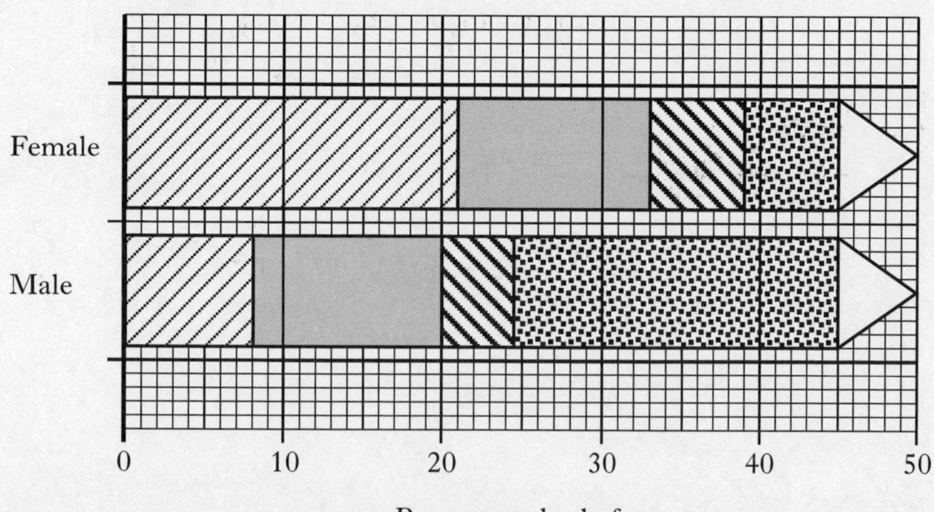

(i) What term would be used to describe a male in this age group whose body fat is 35%?

____ **1**

(ii) What is the **normal** range of percentage body fat for a 20–39 year old female?

Between __________% and __________% **1**

(iii) State **one** conclusion which can be drawn from the information in the chart.

____ **1**

(*b*) State **one** health problem which could result from being overweight.

____ **1**

[*END OF QUESTION PAPER*]

[Turn over

DO NOT WRITE IN THIS MARGIN

SPACE FOR ANSWERS

ADDITIONAL PIE CHART FOR QUESTION 1(*b*)(i)

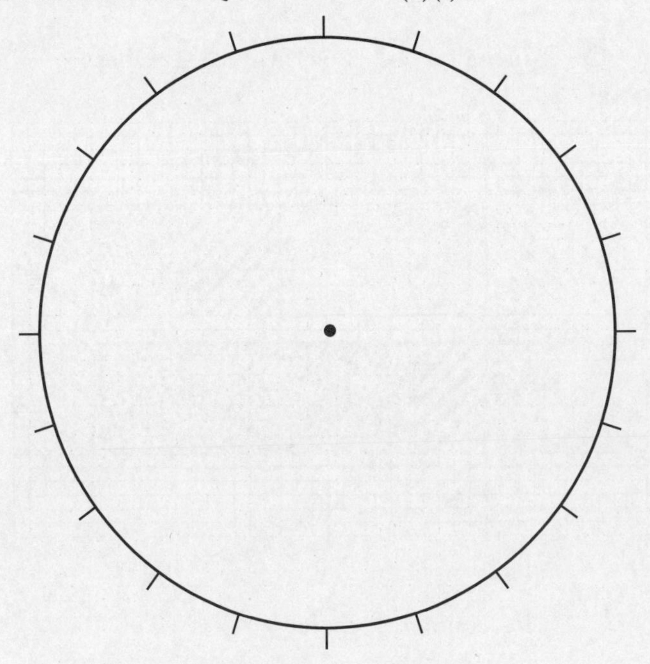

ADDITIONAL GRAPH PAPER FOR QUESTION 2(*a*)(ii)

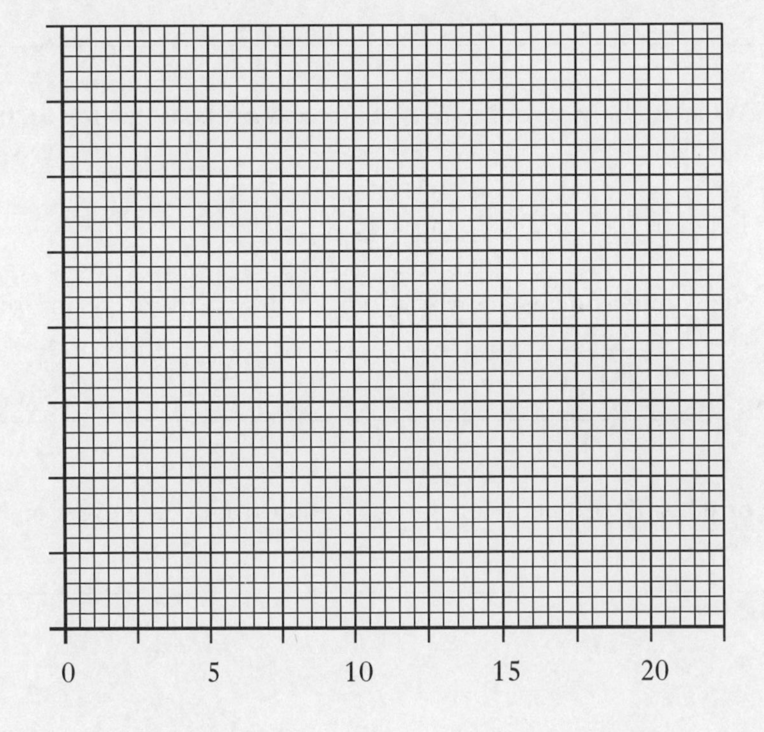

DO NOT WRITE IN THIS MARGIN

SPACE FOR ANSWERS

ADDITIONAL GRAPH PAPER FOR QUESTION 5(*a*)

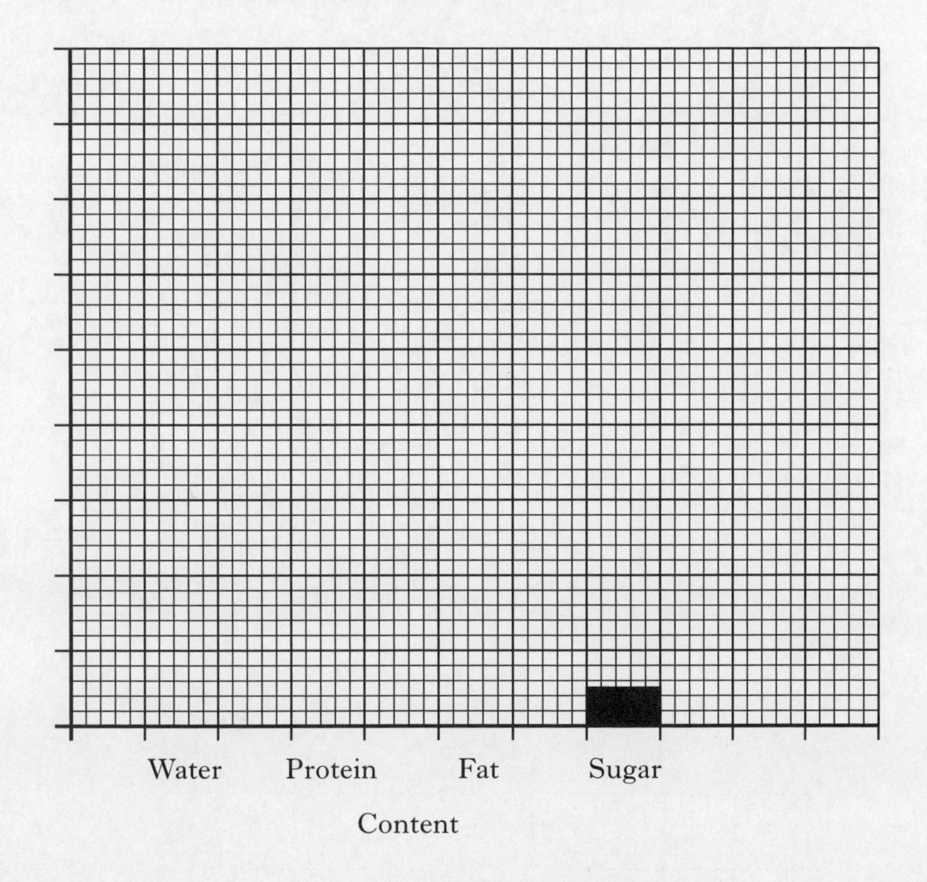

DO NOT WRITE IN THIS MARGIN

SPACE FOR ANSWERS

DO NOT WRITE IN THIS MARGIN

SPACE FOR ANSWERS

DO NOT WRITE IN THIS MARGIN

SPACE FOR ANSWERS

[BLANK PAGE]

FOR OFFICIAL USE

Section B Total

X007/101

NATIONAL QUALIFICATIONS 2007

MONDAY, 21 MAY
9.00 AM – 10.30 AM

BIOLOGY INTERMEDIATE 1

Fill in these boxes and read what is printed below.

Full name of centre

Town

Forename(s)

Surname

Date of birth
Day Month Year

Scottish candidate number

Number of seat

SECTION A

Instructions for completion of Section A are given on page two.

For this section of the examination you must use an **HB pencil**.

SECTION B

1 All questions should be attempted.
2 The questions may be answered in any order but all answers are to be written in the spaces provided in this answer book, **and must be written clearly and legibly in ink**.
3 Additional space for answers will be found at the end of the book. If further space is required, supplementary sheets may be obtained from the invigilator and should be inserted inside the **front** cover of this book.
4 The numbers of questions must be clearly inserted with any answers written in the additional space.
5 Rough work, if any should be necessary, should be written in this book and then scored through when the fair copy has been written. If further space is required, a supplementary sheet for rough work may be obtained from the invigilator.
6 Before leaving the examination room you must give this book to the invigilator. If you do not, you may lose all the marks for this paper.

Read carefully

1 Check that the answer sheet provided is for **Biology Intermediate 1 (Section A)**.

2 For this section of the examination you must use an **HB pencil** and, where necessary, an eraser.

3 Check that the answer sheet you have been given has **your name**, **date of birth**, **SCN** (Scottish Candidate Number) and **Centre Name** printed on it.

Do not change any of these details.

4 If any of this information is wrong, tell the Invigilator immediately.

5 If this information is correct, **print** your name and seat number in the boxes provided.

6 The answer to each question is **either** A, B, C or D. Decide what your answer is, then, using your pencil, put a horizontal line in the space provided (see sample question below).

7 There is **only one correct** answer to each question.

8 Any rough working should be done on the question paper or the rough working sheet, **not** on your answer sheet.

9 At the end of the exam, put the **answer sheet for Section A inside the front cover of this answer book**.

Sample Question

Which of the following foods contains a high proportion of fat?

A Butter

B Bread

C Sugar

D Apple

The correct answer is **A**—Butter. The answer **A** has been clearly marked in **pencil** with a horizontal line (see below).

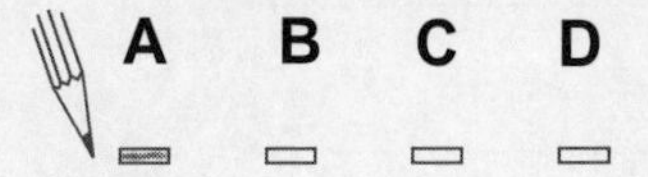

Changing an answer

If you decide to change your answer, carefully erase your first answer and using your pencil, fill in the answer you want. The answer below has been changed to **D**.

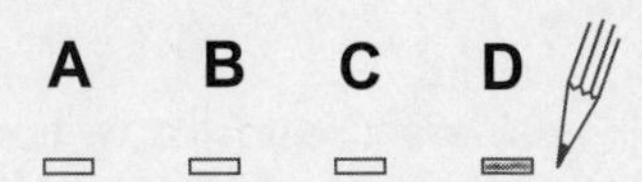

SECTION A

All questions in this section should be attempted.

Answers should be given on the separate answer sheet provided.

1. The process of food production in a leaf is called

A photosynthesis

B propagation

C germination

D fermentation.

2. The diagram shows plants growing in a hanging basket.

One way of preventing compost in a hanging basket from drying out in hot weather is to use

A capillary matting

B grit

C water retentive gel

D sand.

3. An advantage of dormancy is that

A seeds are able to germinate in winter

B seeds require less water for germination

C seed germination is delayed until spring

D seeds germinate more quickly.

[Turn over

4. The steps for taking stem cuttings are shown below.

1 Remove the lower leaves.

2 Place the cuttings into compost.

3 Cut the stem below a node.

4 Cover the pot with a polythene bag.

5 Dip the cut end into rooting powder.

Which of the following shows the correct sequence of steps?

A $3 \rightarrow 5 \rightarrow 4 \rightarrow 2 \rightarrow 1$

B $3 \rightarrow 1 \rightarrow 5 \rightarrow 2 \rightarrow 4$

C $3 \rightarrow 4 \rightarrow 5 \rightarrow 1 \rightarrow 2$

D $3 \rightarrow 5 \rightarrow 2 \rightarrow 4 \rightarrow 1$

5. Grey mould on strawberry plants can be controlled using

A soapy water

B fertiliser

C fungicide

D insecticide.

6. The diagram shows a Mother-in-law's tongue plant being propagated.

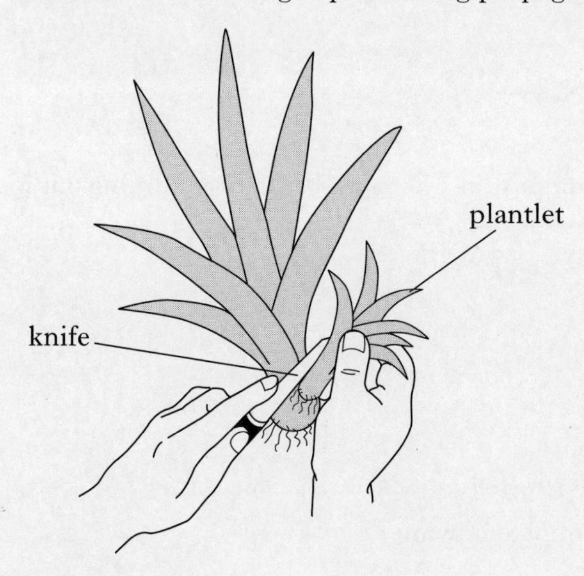

The type of plantlet being removed from the base of the parent plant is

A a bulb

B a tuber

C an offset

D a runner.

7. A student carried out an investigation into the water content of pea seeds.
A pot of fresh seeds was weighed and then placed in an oven at 100 °C until there was no more loss in mass.

The results are shown in the table below.

Condition of peas	*Mass* (g)
Fresh	200
Dry	20

To improve the **reliability** of the results, the student should

A repeat the experiment

B set the oven at 150 °C

C leave the peas in the oven for longer

D use a smaller number of peas.

8. A student carried out an investigation into root development in cuttings.

Six pots were set up as shown in the diagram.

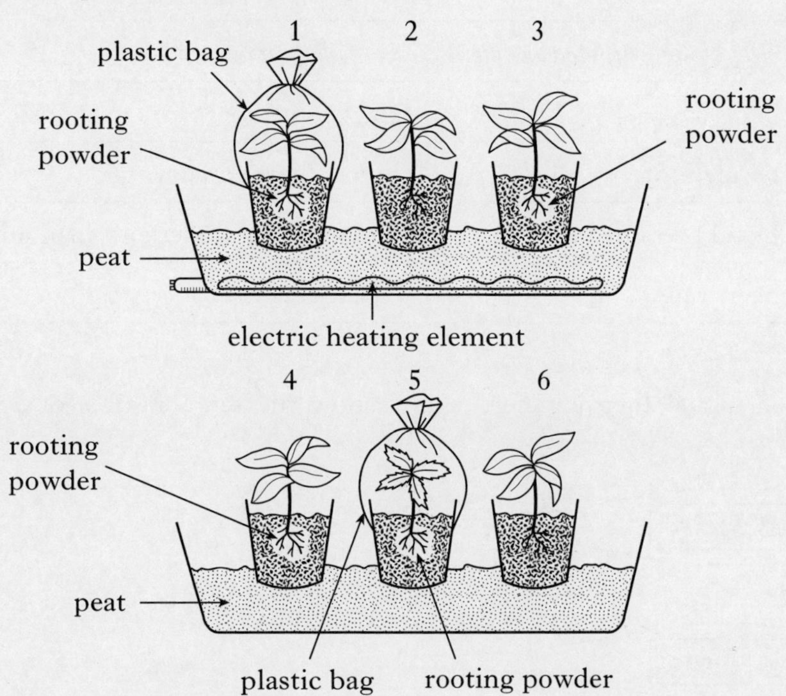

Which pots should be compared to investigate the effect of increasing humidity on the root development in cuttings?

A 1 and 3

B 1 and 4

C 2 and 4

D 3 and 6

9. The table below shows how often a sample of Scottish students take exercise.

Frequency of exercise	*Number of students*	
	Male	*Female*
Daily	35	20
2–3 times a week	25	30
Once a week	7	18
Once a month	3	9

From the table, which of the following statements is correct?

A Fewer females exercise once a month than males.

B More males exercise once a week than females.

C Fewer females exercise daily than males.

D More males exercise 2–3 times a week than females.

10. Which line in the table below correctly shows a physiological measurement and a **high-tech** instrument used to measure it?

	Physiological Measurement	*Instrument*
A	body fat	skin fold callipers
B	temperature	clinical thermometer
C	blood pressure	stethoscope and mercury manometer
D	heart rate	pulsometer

11. The diagrams of digital thermometers below show the temperatures of four patients in a hospital.

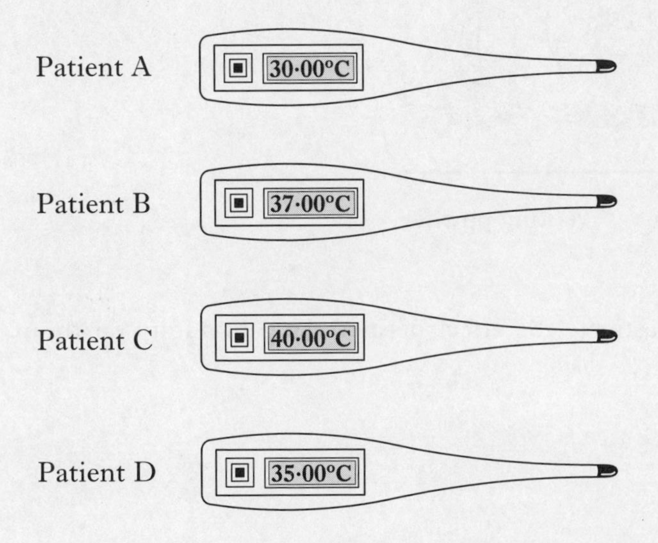

Which patient is most likely to be suffering from a fever?

12. Diabetes can be detected by measuring the blood level of

A antibodies

B sugar

C iron

D white blood cells.

13. The bar graph below shows the percentage of men in different age groups who are smokers.

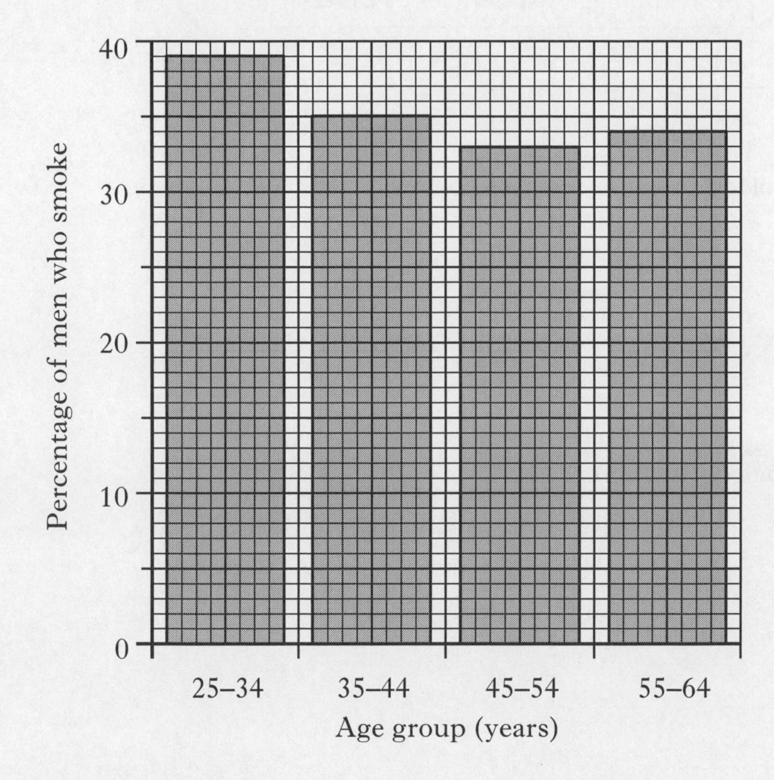

What percentage of 25–34 year old men smoke?

A 33

B 34

C 35

D 39

[Turn over

14. Which line in the table below correctly matches the food group to its use?

	Food group	*Use*
A	fats	growth and repair of cells
B	carbohydrates	energy
C	proteins	protection against deficiency disease
D	vitamins	energy

15. Some of the factors which affect blood pressure are listed below.

Factors:

1 lack of exercise

2 balanced diet

3 being overweight

4 stress

Which of these can lead to high blood pressure?

A Factors 1 and 2 only

B Factors 1 and 4 only

C Factors 1, 2 and 3 only

D Factors 1, 3 and 4 only

16. The table below shows the components which make up the body mass of an adult female.

Component	*Percentage of adult female's body mass*
Fat	40
Muscle	25
Bone	20
Other tissues	15

Which of the following pie charts presents this information correctly?

Key
- Fat
- Muscle
- Bone
- Other tissues

A

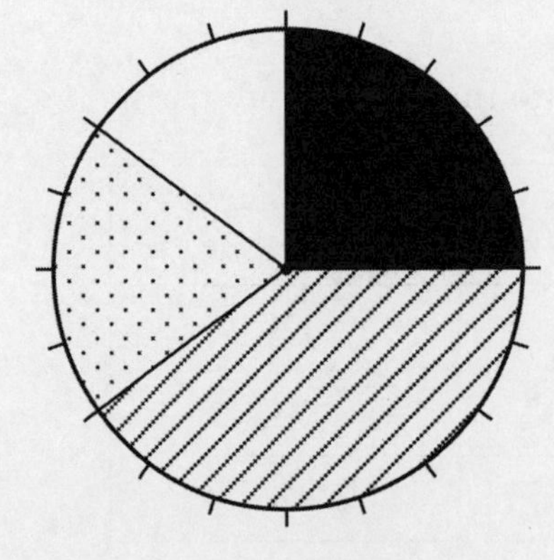

B

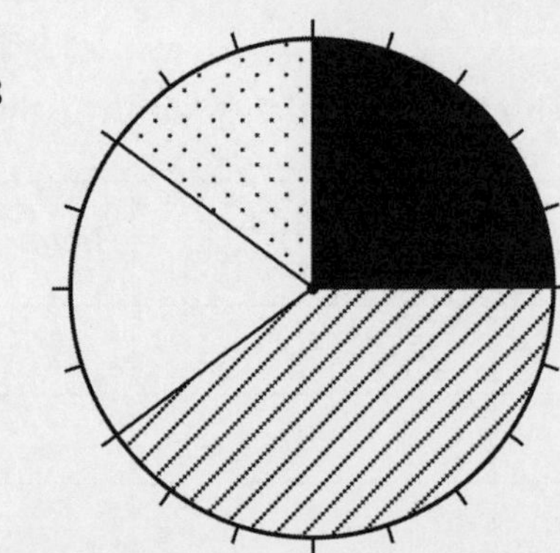

C

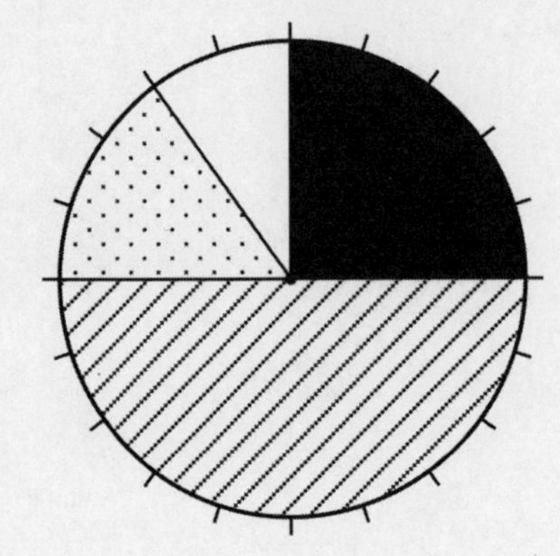

D

[Turn over

17. The steps in an investigation into the fitness of two students are outlined below.

1. Both students measured their resting pulse rate using a pulsometer.
2. One student used an exercise bike for ten minutes while the other ran for the same length of time.
3. Both students measured their pulse rate after the exercise for twenty minutes.
4. The investigation was repeated.

Which of the following is a possible source of error in this investigation?

A A pulsometer was used.

B The period of exercise was only ten minutes.

C The investigation was repeated.

D The students performed different exercises.

18. Heating milk to remove some liquid makes it more concentrated.

The type of milk produced by this treatment is

A evaporated

B skimmed

C semi-skimmed

D UHT.

19. The table below shows the results of an investigation into the removal of stains.

Type of stain	*Washing temperature* (°C)	*Biological detergent*	*Non-biological detergent*
Grass	40	✓	×
Mud	40	✓	×
Grass	100	×	×
Mud	100	✓	✓

✓ = stain removed × = stain not removed

Grass stains were removed by a

A biological detergent at 40 °C

B biological detergent at 100 °C

C non-biological detergent at 40 °C

D non-biological detergent at 100 °C.

20. Waste whey can be upgraded to produce

A cheese and animal feed

B animal feed and a creamy alcoholic drink

C yoghurt and a creamy alcoholic drink

D cheese and yoghurt.

21. Which of the following is added to the food of farmed salmon to give their flesh the same appearance as wild salmon?

A Flavouring

B Minerals

C Dried whey

D Yeast products

22. The table below shows the temperature ranges in which different types of yeast can grow.

Type of yeast	*Temperature range* (°C)
1	14–16
2	12–30
3	18–35
4	8–15

Which types of yeast will **not** grow at 16 °C?

A Types 1 and 3

B Types 1 and 4

C Types 2 and 3

D Types 3 and 4

23. Which of the following is added to milk to make yoghurt?

A Antibiotics

B Bacteria

C Enzymes

D Yeast

[Turn over

24. Which line in the table below correctly identifies the effects of discharging whey into a river?

	Number of bacteria	*Oxygen availability*
A	increase	increase
B	increase	decrease
C	decrease	increase
D	decrease	decrease

25. The key below shows different types of infection and their treatment.

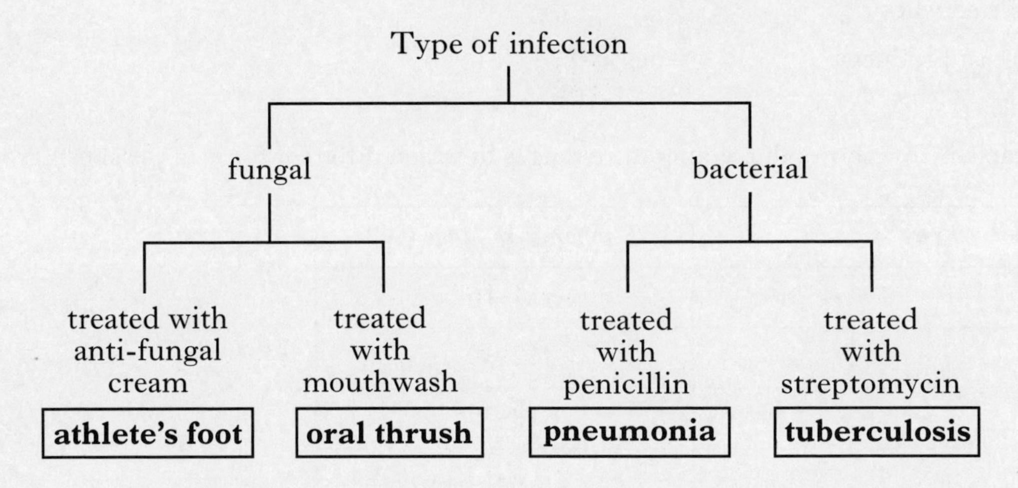

Which of the following correctly identifies tuberculosis from this information?

	Type of infection	*Treatment*
A	bacterial	streptomycin
B	fungal	anti-fungal cream
C	bacterial	penicillin
D	fungal	mouthwash

Candidates are reminded that the answer sheet for Section A MUST be returned <u>inside</u> this answer book.

DO NOT WRITE IN THIS MARGIN

Marks

SECTION B

All questions in this Section should be attempted.
All answers must be written clearly and legibly in ink.

1. Read the following passage carefully.

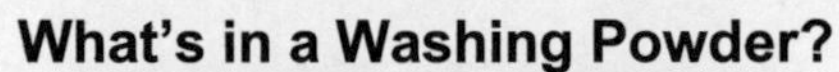

Modern washing powders contain a number of chemicals which reflect the complex demands of modern living. These detergents must remove stains without damaging fabrics and washing machines. They should also be environmentally friendly.

Most detergents contain surfactants which allow the water to spread across the fabric and builders to soften the water. In addition, lather control agents are put in to stop too much froth forming. The pleasant smell of detergents is produced by fragrances. Corrosion inhibitors protect the washing machine from rusting.

Biological washing powders also contain several types of enzyme such as proteases, amylases and lipases. These enzymes are so powerful that the powders have only 1% enzymes.

Answer the questions below, using information from the passage.

(*a*) Name the chemical in washing powders which prevents too much froth forming.

________________________________ **1**

(*b*) Why are "builders" added to washing powders?

____ **1**

(*c*) Why do biological washing powders contain a very low percentage of enzymes?

____ **1**

(*d*) Name an enzyme found in biological washing powders.

________________________________ **1**

DO NOT WRITE IN THIS MARGIN

Marks

2. The following flow chart shows some of the processes involved in cheese making.

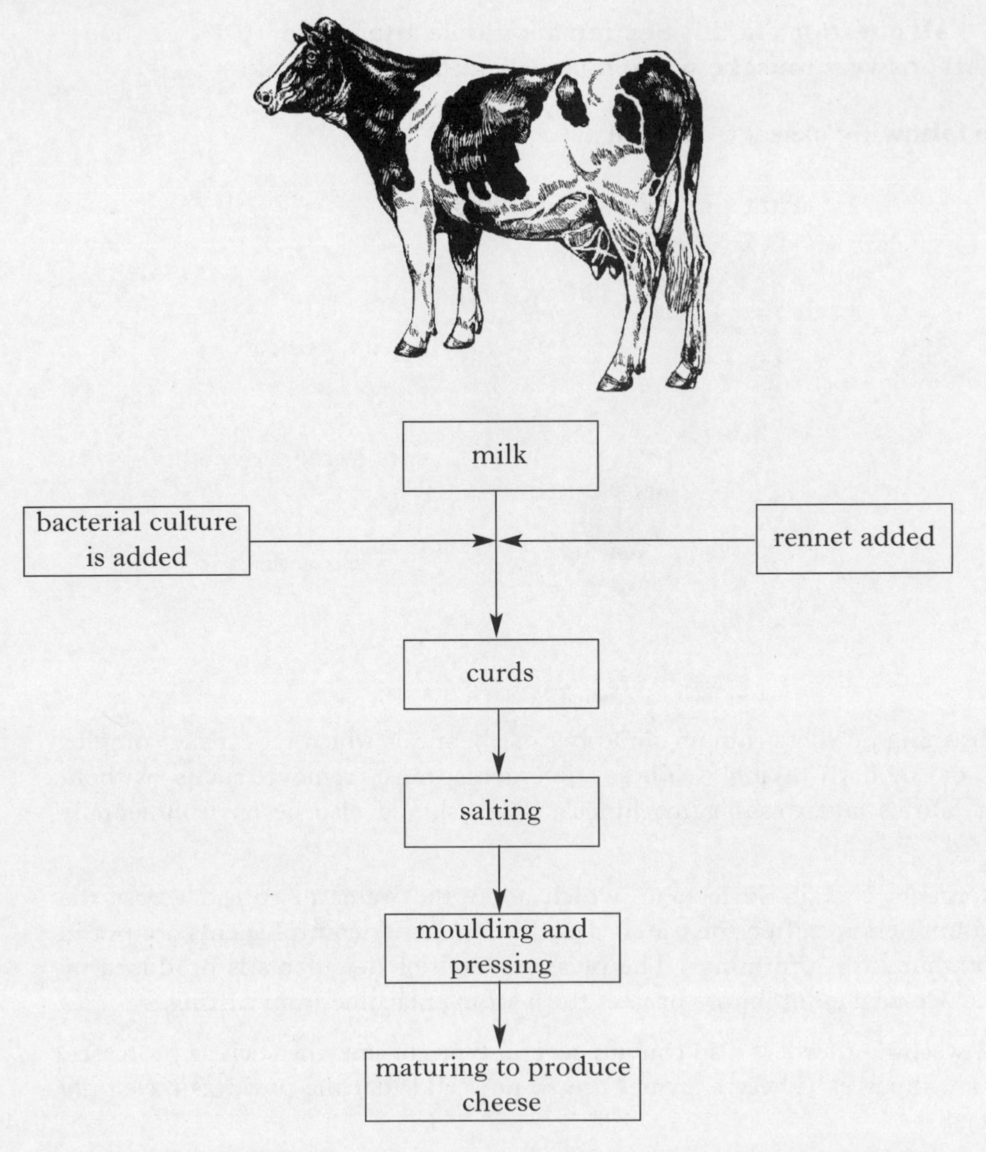

(*a*) State **one** source of the rennet which is used in cheese making. **1**

____

(*b*) Name the liquid which is left behind when the curds are formed. **1**

____

(*c*) **Underline** one option in each set of brackets to make the sentences below correct.

Bacterial cultures are added to milk to help convert {sugar / acid} into {acid / sugar}.

This helps to clot the {protein / fat} and also affects the {colour / flavour} of the cheese. **2**

DO NOT WRITE IN THIS MARGIN

Marks

3. (*a*) The steps in an investigation into the effect of six antibiotics on bacterial growth are shown below.

Step 1 An agar plate was evenly spread with a type of bacteria.

Step 2 Six different antibiotic discs (A–F) were placed on the agar.

Step 3 The plate was left at 25 °C for 48 hours.

Step 4 The size of any clear zone around each disc was measured.

The results are shown below.

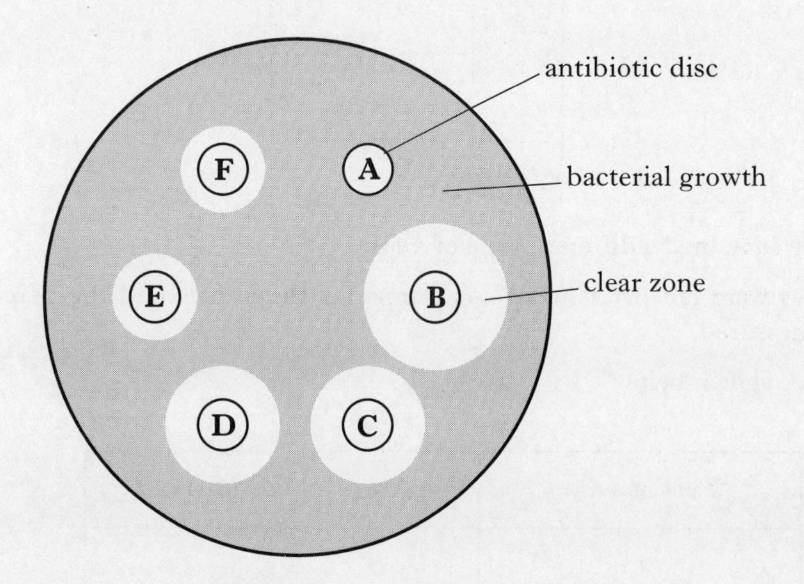

(i) The conclusion below was written by the student who carried out the investigation.

> **Conclusion:**
> Antibiotic A is not effective against any type of bacteria

Explain why this is **not** a valid conclusion.

____

____ **1**

(ii) Describe **one** way in which this investigation could be improved.

____ **1**

(*b*) Why have some bacteria developed resistance to antibiotics?

____ **1**

[Turn over

DO NOT WRITE IN THIS MARGIN

Marks

4. A student carried out an investigation into the production of alcohol by yeast.

Four test tubes each containing water, yeast and sugar were set up as shown.

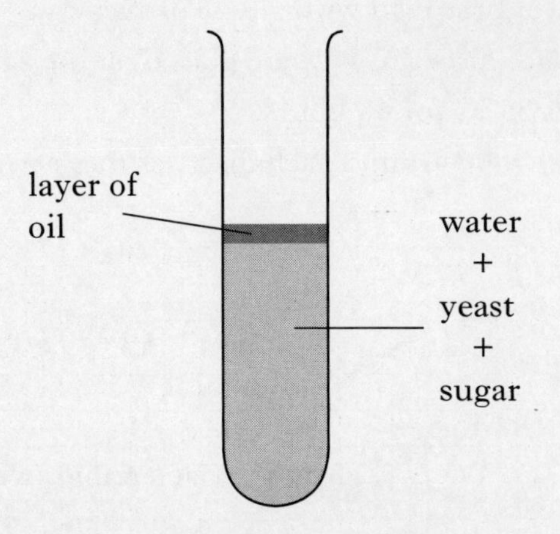

Each test tube contained a different type of yeast.

The test tubes were left in identical conditions for three days and the percentage alcohol was measured.

The results are shown below.

Type of yeast	*Percentage alcohol after 3 days*
1	2·4
2	3·4
3	4·0
4	3·0

(*a*) (i) On the grid opposite, complete the **bar chart** by

(1) providing a label on the vertical axis **1**

(2) putting a scale on the vertical axis **1**

(3) plotting the remaining results. **1**

(Additional graph paper, if required, will be found on page 28.)

DO NOT WRITE IN THIS MARGIN

Marks

4. (*a*) (i) (continued)

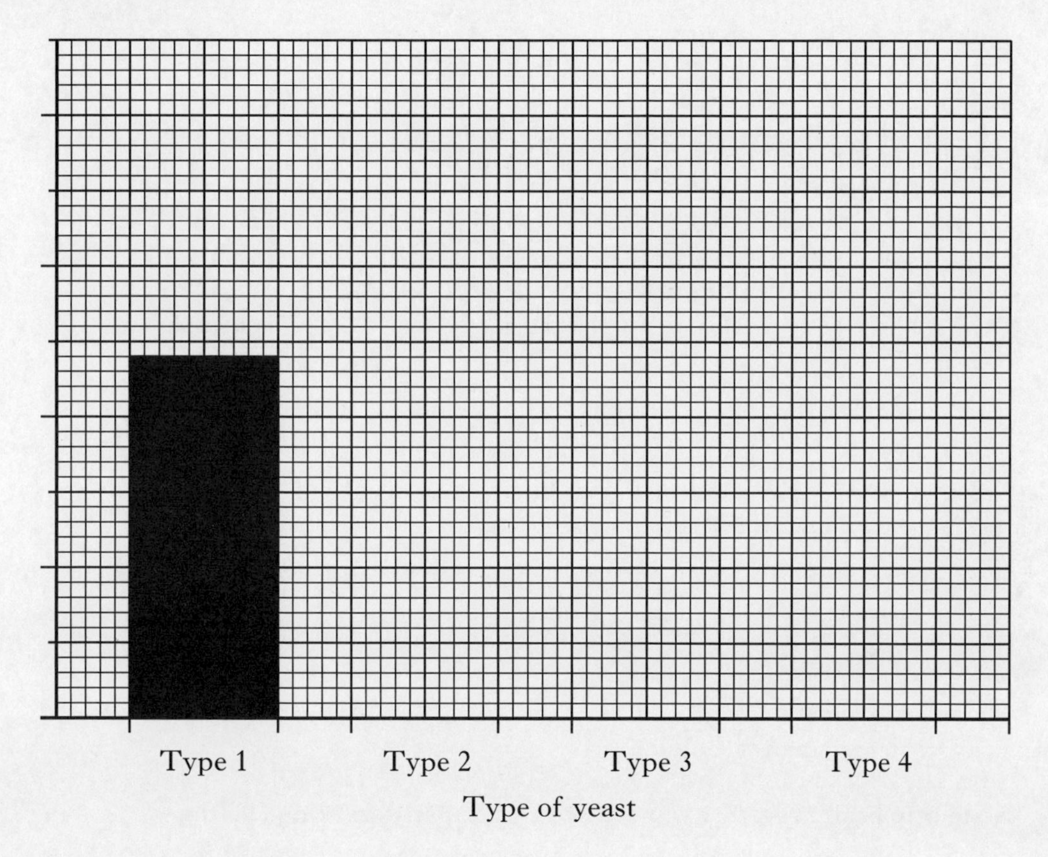

(ii) Draw **one** conclusion from the results.

____

______________________________ **1**

(*b*) (i) Name the process by which alcohol is produced by yeast in beer making.

______________________________ **1**

(ii) Name another substance produced in this process.

______________________________ **1**

[Turn over

DO NOT WRITE IN THIS MARGIN

Marks

5. The mass of tar in four different brands of cigarette is shown below.

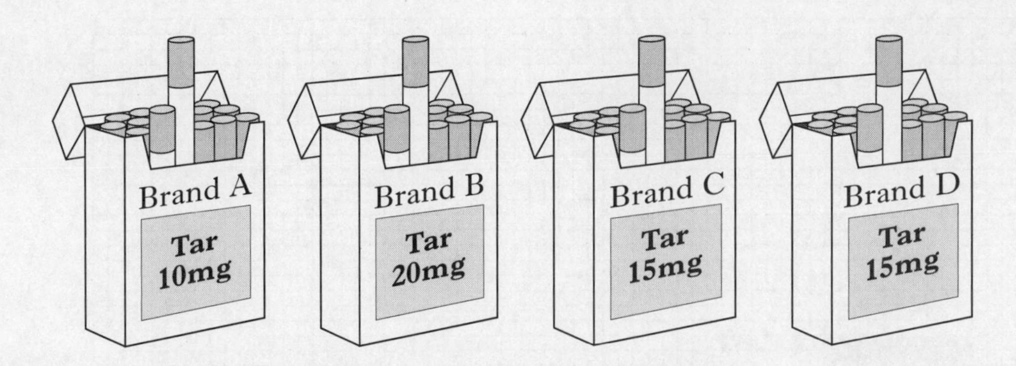

(*a*) Calculate the average mass of tar in the four brands of cigarette.

Space for calculation

________________ mg **1**

(*b*) State **one** health risk that is increased by regularly smoking cigarettes.

____ **1**

(*c*) Cigarette smoke also contains carbon monoxide.

Describe the effect carbon monoxide has on the ability of the blood to carry oxygen.

____ **1**

DO NOT WRITE IN THIS MARGIN

Marks

6. The table below shows the percentage composition of a veggie burger.

Component	*Composition* (%)
Protein	50
Fat	15
Fibre	25
Water	10

(*a*) Present the information in the table in the form of a pie chart.

(An additional pie chart, if required, will be found on page 28.)

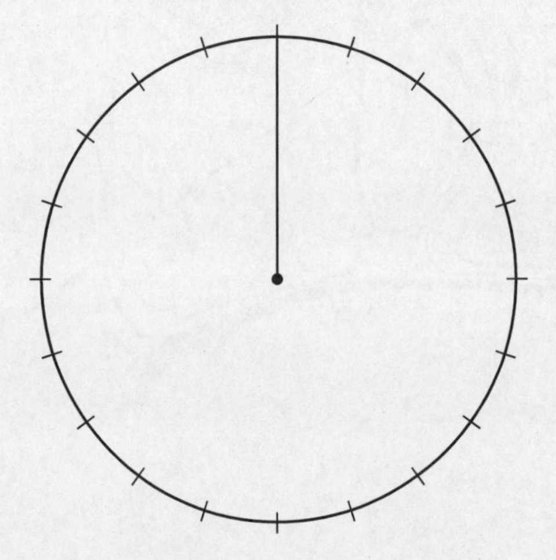

2

(*b*) The veggie burger weighs 50 grams.
Calculate the mass of protein present in the veggie burger.
Space for calculation

________________ grams **1**

(*c*) A beef burger contains 35% fat.

Calculate the simple whole number ratio of fat in a beef burger to fat in a veggie burger.
Space for calculation

________________ : ________________ **1**

beef burger veggie burger

DO NOT WRITE IN THIS MARGIN

Marks

7. A diagram of part of the breathing system is shown below.

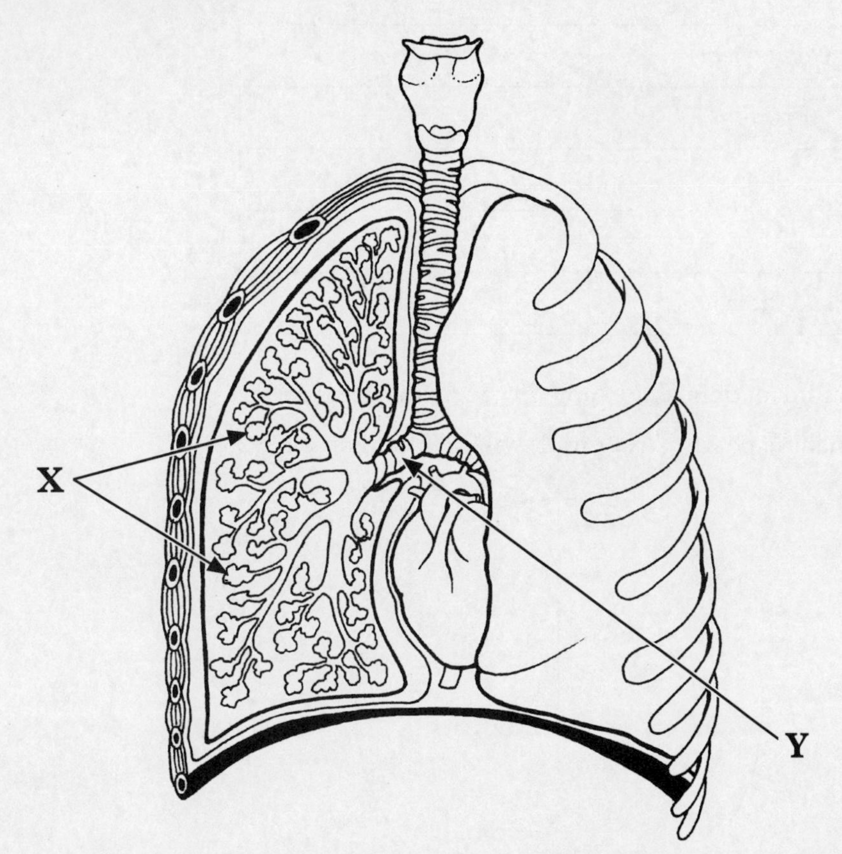

(*a*) (i) Name structure Y.

______________________________ **1**

(ii) Name the gas that is **removed from** the blood at structure X.

______________________________ **1**

(*b*) Draw arrows to link each physiological measurement with its correct definition.

Physiological measurement	*Definition*
Tidal volume	Maximum volume of air breathed out in one breath after breathing in as deeply as possible
Vital capacity	Maximum rate at which air can be forced out of the lungs
Peak flow	Volume of air breathed in or out in one normal breath

2

DO NOT WRITE IN THIS MARGIN

Marks

8. (*a*) The table below shows the blood group of a number of students.

Blood group	*Number of students*
A	4
B	4
O	11
AB	1

Calculate the percentage of students with blood group O.
Space for calculation

__________ % **1**

(*b*) The diagram represents part of the circulatory system.

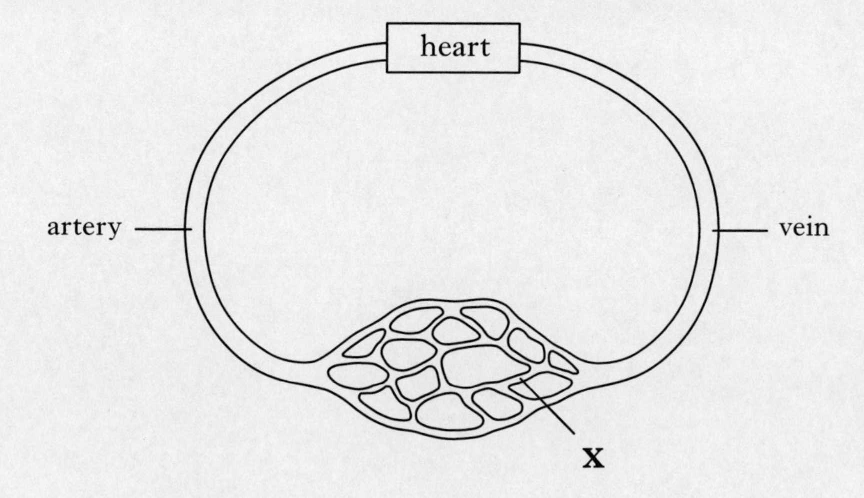

(i) Name the type of blood vessel labelled X.

____ **1**

(ii) Draw arrows on the diagram to show the direction of blood flow in the artery and the vein. **1**

(iii) What is the function of the heart?

____ **1**

DO NOT WRITE IN THIS MARGIN

Marks

9. (*a*) The diagram shows a germinating kidney bean seed.

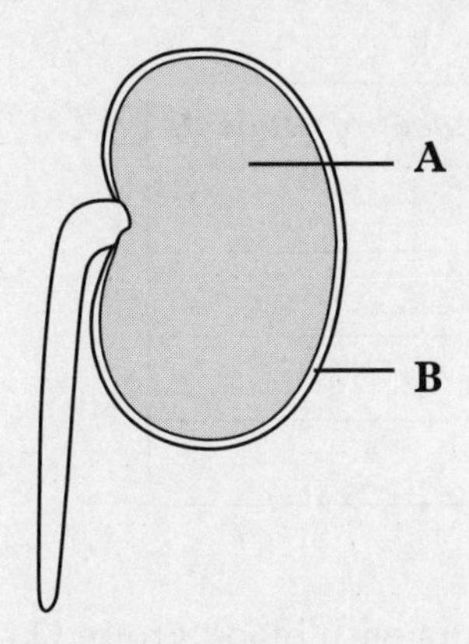

Use the diagram to complete the following table.

Label	*Name*	*Function*
A		Provides energy for growth
B	Seed coat	

2

DO NOT WRITE IN THIS MARGIN

Marks

9. (continued)

(*b*) The root length of the germinating kidney bean seedling was measured every two days.

The results are shown in the table below.

Time (days)	*Root length* (mm)
0	0
2	4
4	8
6	18
8	27

(i) On the grid below, complete the **line graph** by

(1) providing a label for the horizontal axis **1**

(2) completing the scale on the vertical axis **1**

(3) plotting the results. **1**

(Additional graph paper, if required, will be found on page 29.)

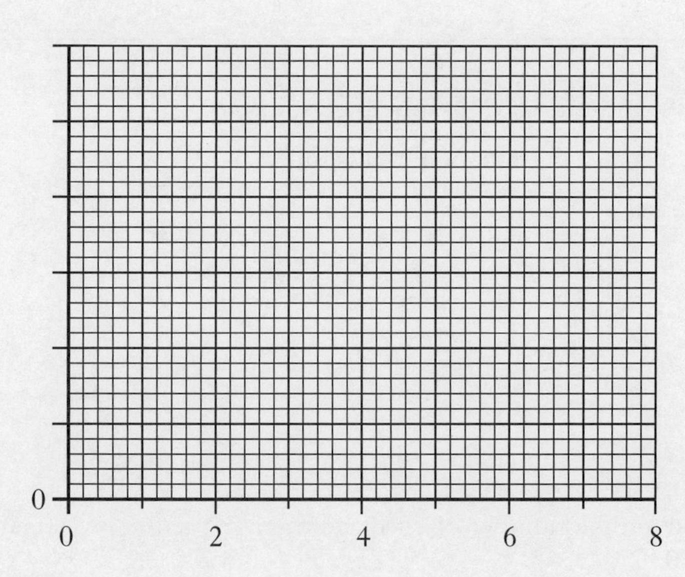

(ii) Between which **two** days was there the greatest increase in root length?

Between day ________ and day ________ . **1**

DO NOT WRITE IN THIS MARGIN

Marks

10. (*a*) An investigation into the conditions required for germination was carried out as shown below.

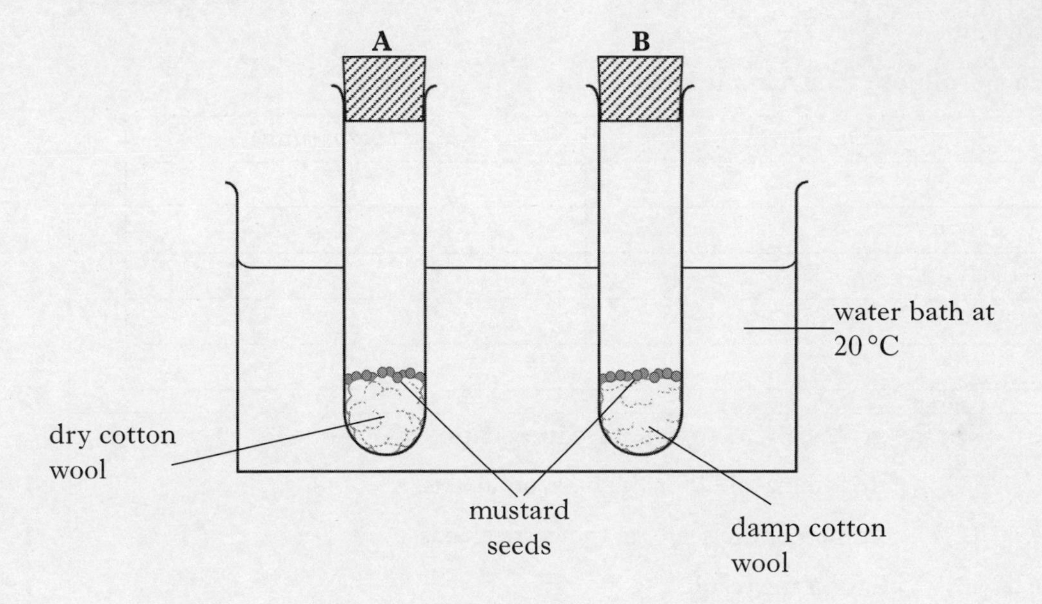

(i) Which variable is being investigated?

____ **1**

(ii) Twelve of the fifteen seeds sown in tube B germinated.

Calculate the percentage germination in this tube.

Space for calculation

____________% **1**

(iii) Predict the effect on the number of seeds germinating if the investigation was repeated at 0 °C.

____ **1**

(*b*) Some seeds have thick seed coats which are split open before sowing.

What name is given to this process?

____ **1**

DO NOT WRITE IN THIS MARGIN

Marks

10. (continued)

(*c*) Various methods are used in maintaining plants.

Decide if each description is true or false and tick (✓) the appropriate box.

If the description is **false**, write the correct word or phrase in the correction box to replace the phrase underlined in the definition.

Definition	*True*	*False*	*Correction*
Dead-heading is the removal of dead flowers to prevent further flowering.			
Pricking out is the removal of seedlings to give more room to grow.			
Potting on is the placing of a growing plant into a larger container.			

3

[Turn over

DO NOT WRITE IN THIS MARGIN

Marks

11. A group of students carried out an investigation into the effect of different concentrations of rooting powder on the root growth of rose cuttings.

Ten cuttings were dipped in different concentrations of rooting powder.

The cuttings were grown in coarse sand for five weeks and the lengths of the roots were measured.

The results are shown below.

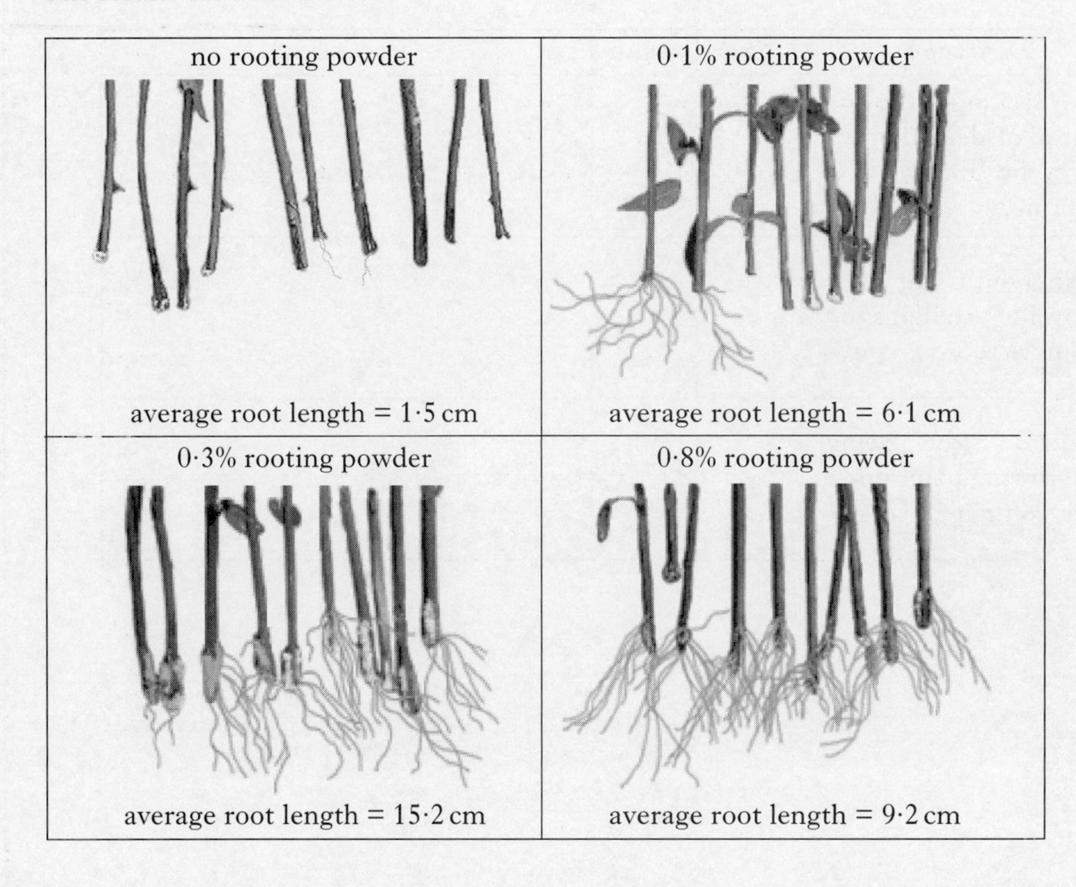

(*a*) Use this information to complete the following table.

Concentration of rooting powder (%)	

2

DO NOT WRITE IN THIS MARGIN

Marks

11. (continued)

(*b*) Explain why cuttings with no rooting powder were grown.

____

____

____ **1**

(*c*) Which concentration of rooting powder had the greatest effect on root length?

__________ % **1**

(*d*) Explain why ten cuttings were used at each concentration of rooting powder.

____

____ **1**

[*END OF QUESTION PAPER*]

DO NOT WRITE IN THIS MARGIN

SPACE FOR ANSWERS

ADDITIONAL GRAPH PAPER FOR QUESTION 4(*a*)(i)

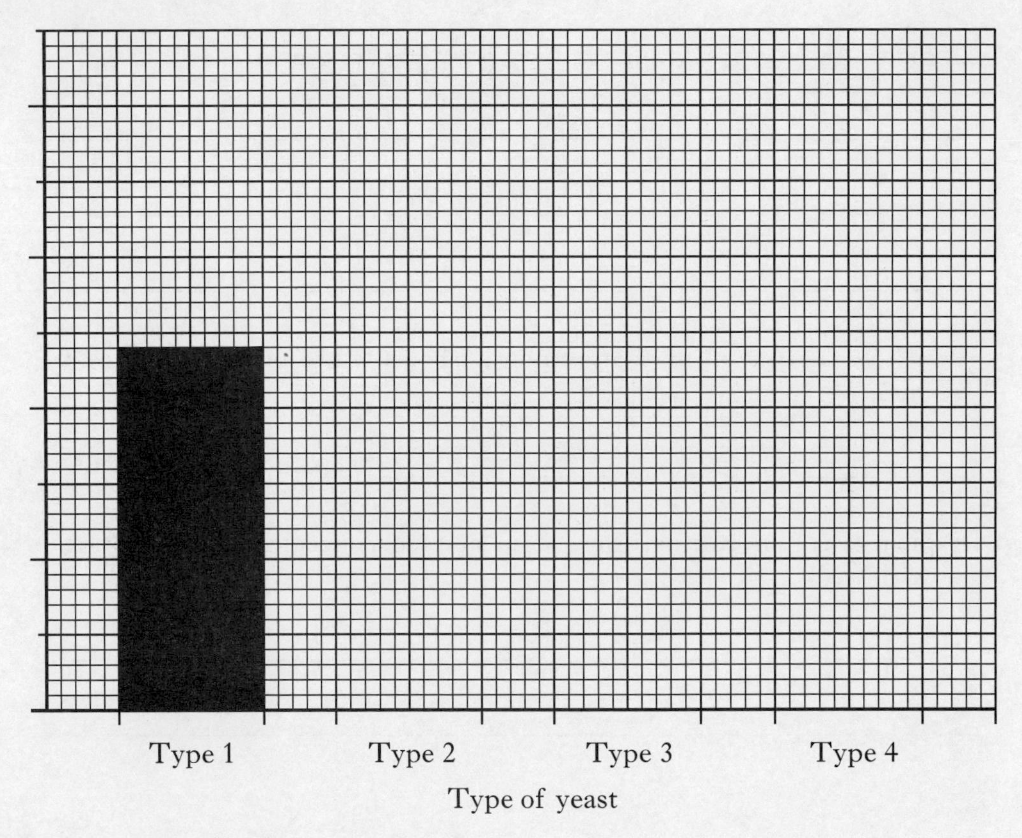

ADDITIONAL PIE CHART FOR QUESTION 6(*a*)

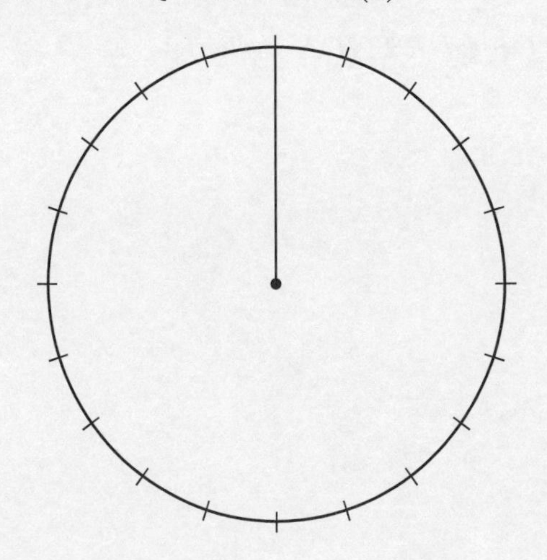

DO NOT WRITE IN THIS MARGIN

SPACE FOR ANSWERS

ADDITIONAL GRAPH PAPER FOR QUESTION 9(*b*)(i)

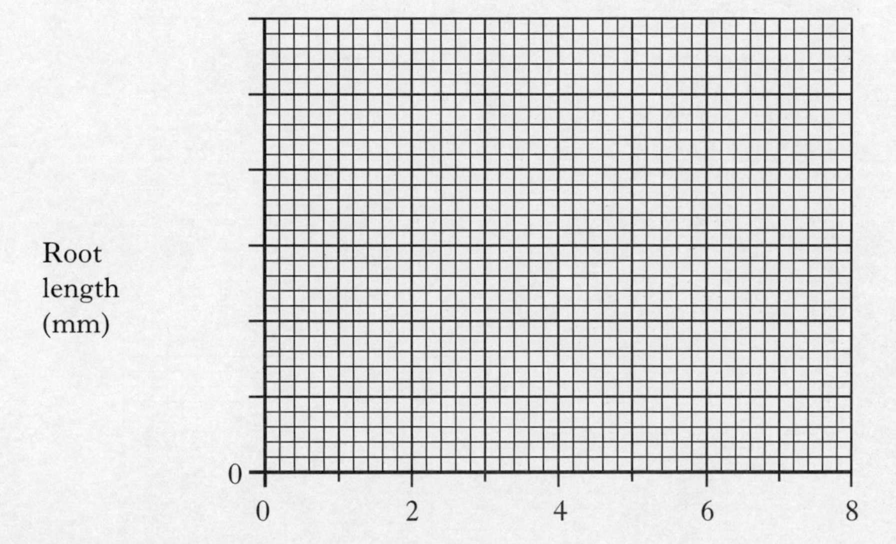

DO NOT WRITE IN THIS MARGIN

SPACE FOR ANSWERS

DO NOT WRITE IN THIS MARGIN

SPACE FOR ANSWERS

DO NOT WRITE IN THIS MARGIN

SPACE FOR ANSWERS

[BLANK PAGE]

FOR OFFICIAL USE

Section B Total

X007/101

NATIONAL QUALIFICATIONS 2008

TUESDAY, 27 MAY 9.00 AM – 10.30 AM

BIOLOGY INTERMEDIATE 1

Fill in these boxes and read what is printed below.

Full name of centre

Town

Forename(s)

Surname

Date of birth

Day Month Year

Scottish candidate number

Number of seat

SECTION A

Instructions for completion of Section A are given on page two.

For this section of the examination you must use an **HB pencil**.

SECTION B

1 All questions should be attempted.

2 The questions may be answered in any order but all answers are to be written in the spaces provided in this answer book, **and must be written clearly and legibly in ink**.

3 Additional space for answers will be found at the end of the book. If further space is required, supplementary sheets may be obtained from the invigilator and should be inserted inside the **front** cover of this book.

4 The numbers of questions must be clearly inserted with any answers written in the additional space.

5 Rough work, if any should be necessary, should be written in this book and then scored through when the fair copy has been written. If further space is required, a supplementary sheet for rough work may be obtained from the invigilator.

6 Before leaving the examination room you must give this book to the invigilator. If you do not, you may lose all the marks for this paper.

Read carefully

1 Check that the answer sheet provided is for **Biology Intermediate 1 (Section A)**.

2 For this section of the examination you must use an **HB pencil** and, where necessary, an eraser.

3 Check that the answer sheet you have been given has **your name**, **date of birth**, **SCN** (Scottish Candidate Number) and **Centre Name** printed on it.

Do not change any of these details.

4 If any of this information is wrong, tell the Invigilator immediately.

5 If this information is correct, **print** your name and seat number in the boxes provided.

6 The answer to each question is **either** A, B, C or D. Decide what your answer is, then, using your pencil, put a horizontal line in the space provided (see sample question below).

7 There is **only one correct** answer to each question.

8 Any rough working should be done on the question paper or the rough working sheet, **not** on your answer sheet.

9 At the end of the exam, put the **answer sheet for Section A inside the front cover of this answer book**.

Sample Question

Which of the following foods contains a high proportion of fat?

A Butter

B Bread

C Sugar

D Apple

The correct answer is **A**—Butter. The answer **A** has been clearly marked in **pencil** with a horizontal line (see below).

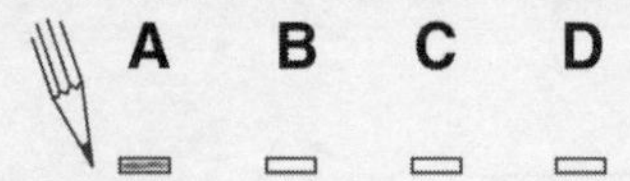

Changing an answer

If you decide to change your answer, carefully erase your first answer and using your pencil, fill in the answer you want. The answer below has been changed to **D**.

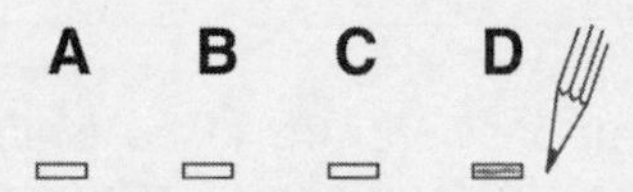

SECTION A

All questions in this section should be attempted.

Answers should be given on the separate answer sheet provided.

1. There are three aspects to health shown below in the health triangle.

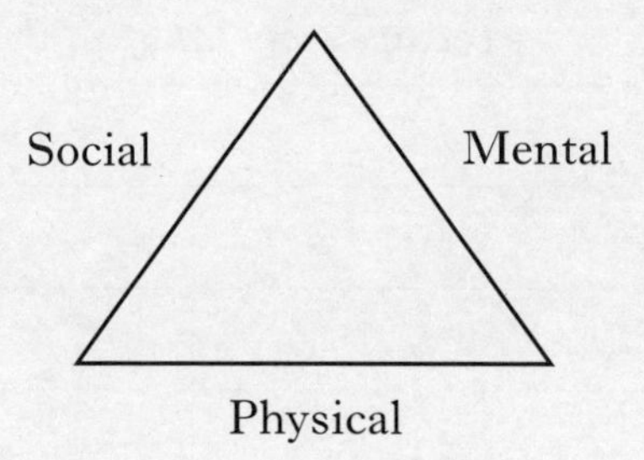

Which of the following is a physical aspect of health?

A No stress at work

B Eating a balanced diet

C Looking forward to the weekend

D Enjoying the company of friends

2. Which line in the table below shows correctly the change in concentrations of oxygen and carbon dioxide in the blood as it passes through the lungs?

	Concentration in blood	
	Oxygen	*Carbon dioxide*
A	increases	decreases
B	increases	increases
C	decreases	decreases
D	decreases	increases

3. Oxygen moves into the tissues as the blood is flowing through the

A capillaries

B arteries

C veins

D arteries and veins.

[Turn over

4. Males have on average between 15% and 17% body fat.
Females have on average between 18% and 22% body fat.

The table below gives average percentage body fat for athletes in four sports.

	Sport	*Average body fat of athletes* (%)	
		Male	*Female*
A	Swimming	10	16
B	Running	9	12
C	Volleyball	11	16
D	Shotput	18	24

In which sport do the athletes have a higher than average percentage of body fat?

5. Which line in the table indicates correctly the increased health risks for being overweight or underweight?

	Overweight	*Underweight*
A	Anorexia	Cancer
B	Diabetes	Heart disease
C	Arthritis	Anorexia
D	Cancer	Arthritis

6. The blood groups of 200 students are shown in the table below.

Blood Group	*Number of Students*
O	94
A	84
B	16
AB	6

What percentage of the students have Blood Group A?

A 42%

B 45%

C 84%

D 90%

7. Which row in the table below describes correctly a health problem linked to blood pressure?

	Health Problem	*Blood Pressure*
A	angina	low
B	heart attack	low
C	fainting	high
D	stroke	high

8. The table and pie chart contain the same information about the diet of British people.

Type of Food	*Percentage of Diet*
Cereals	50
Animal protein	25
Fruit and vegetables	12·5
Others	12·5

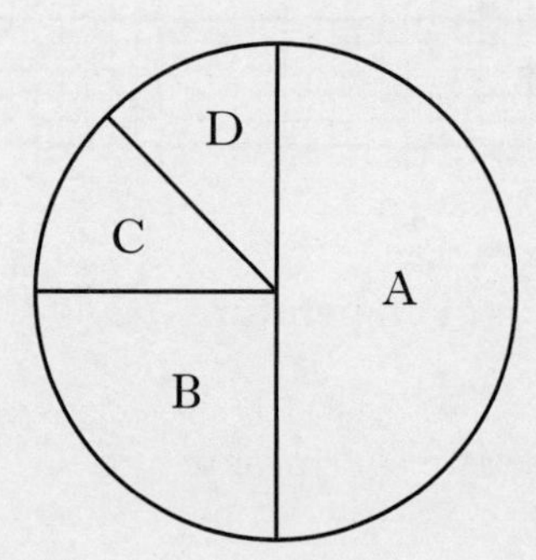

Animal protein is represented by which slice of the pie chart?

9. When sowing fine seeds they should be

A pre-germinated

B spaced out individually

C mixed with silver sand

D mixed with larger seeds.

10. The following graph shows the effect of temperature on the germination of oat seeds and barley seeds.

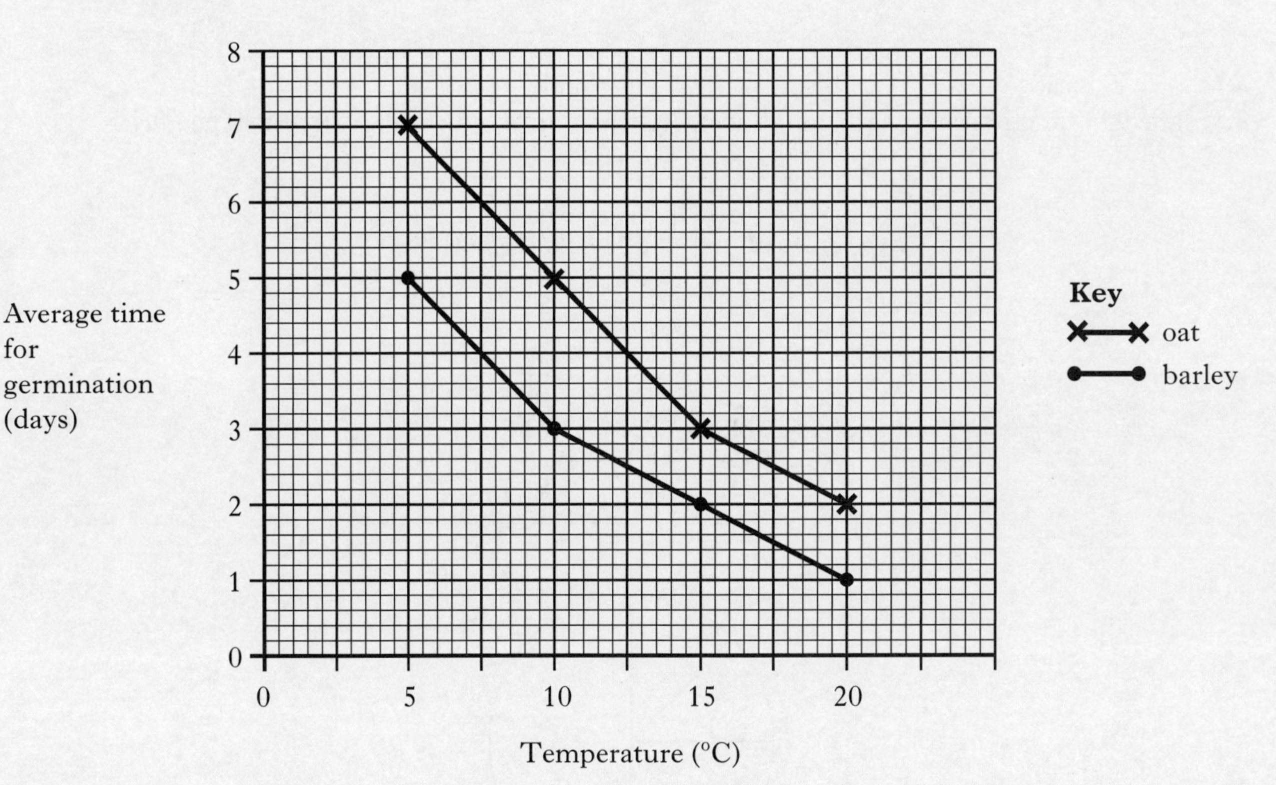

What is the average time for oat seeds to germinate at 5 °C?

A 5 days

B 6 days

C 7 days

D 10 days

11. A student tested four types of seeds for the presence of starch, sugar and protein. The tests used were:

Starch present – iodine solution turns from brown to black
Sugar present – clinistix turns from pink to purple
Protein present – albustix turns from yellow to green

The results are shown in the table below.

Seed type	*Colour produced*		
	Starch test	*Sugar test*	*Protein test*
Barley	black	pink	yellow
Pea	black	pink	green
Cabbage	brown	purple	yellow
Mustard	brown	purple	green

Which type of seed contains only starch?

A Barley

B Pea

C Cabbage

D Mustard

12. The diagram below shows cuttings enclosed in plastic bags.

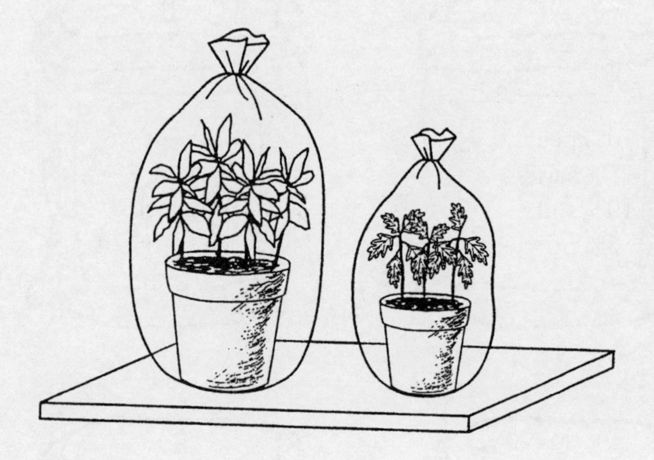

This results in

A an increase in light intensity

B an increase in humidity

C a decrease in temperature

D a decrease in leaf area.

13. Which method of maintaining plants is shown in the diagram below?

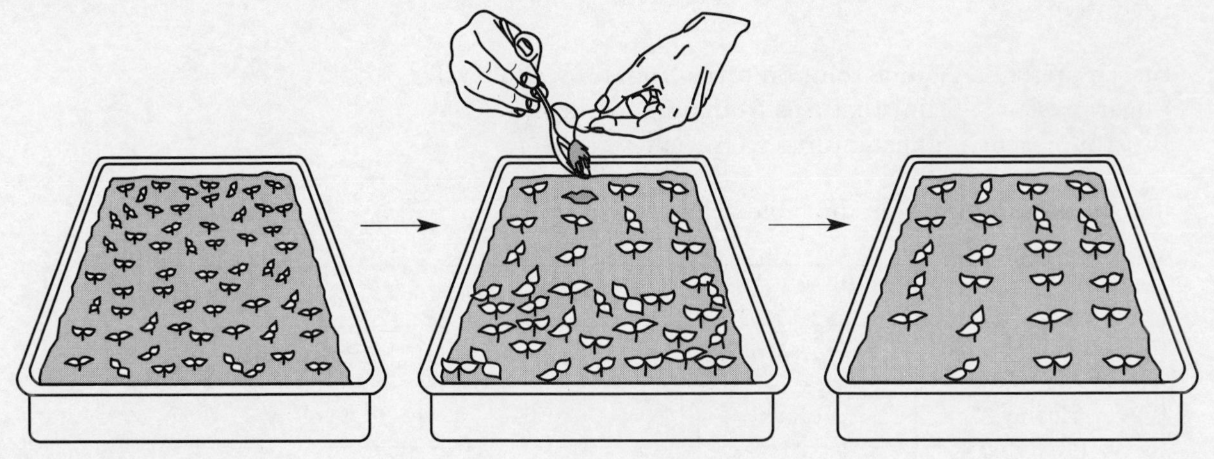

A Layering

B Dead heading

C Pricking out

D Taking cuttings

14. Which of the following is important for good leaf growth?

A Phosphorus

B Nitrogen

C Perlite

D Sand

15. Which of the following composts would have the best drainage?

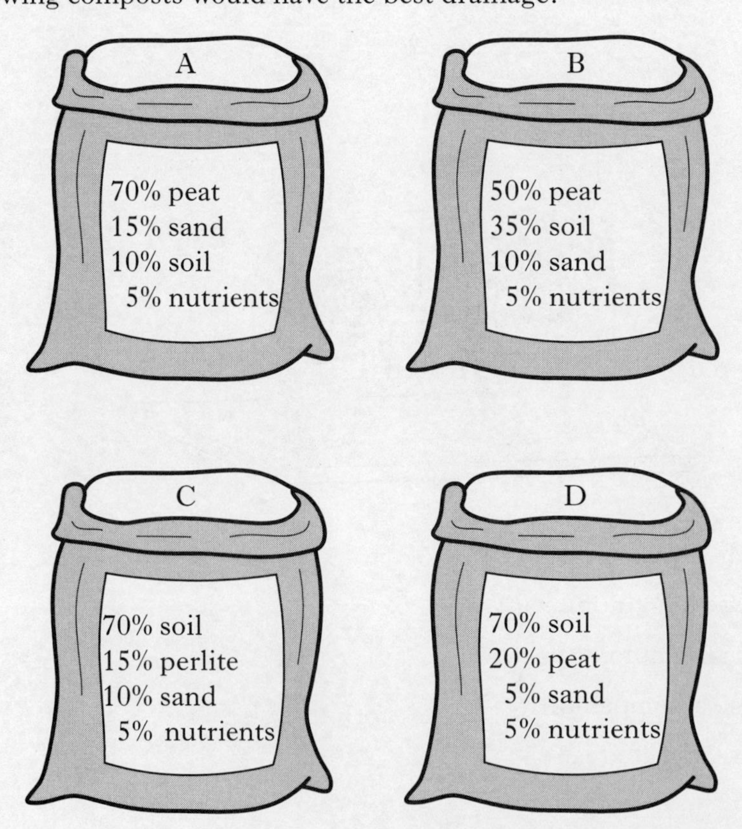

16. The following apparatus was used to investigate the ability of different composts to hold water.

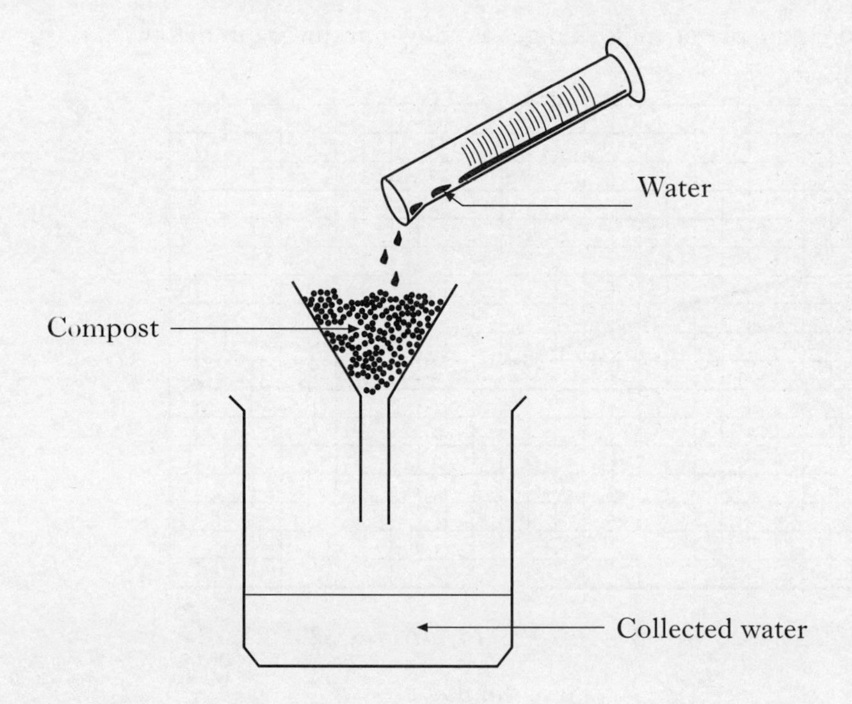

Variables	
1	volume of water poured in
2	mass of compost
3	volume of water collected
4	time taken to collect water

Which variables should be kept constant to allow a valid comparison to be made?

A 1, 3 and 4

B 2, 3 and 4

C 1, 2 and 3

D 1, 2 and 4

17. Which of the following is produced by yeast and makes dough rise?

A Oxygen

B Rennet

C Carbon dioxide

D Alcohol

[Turn over

Questions 18 and **19** refer to the information below.

During yoghurt making, the pH of milk changes as shown in the graph below.

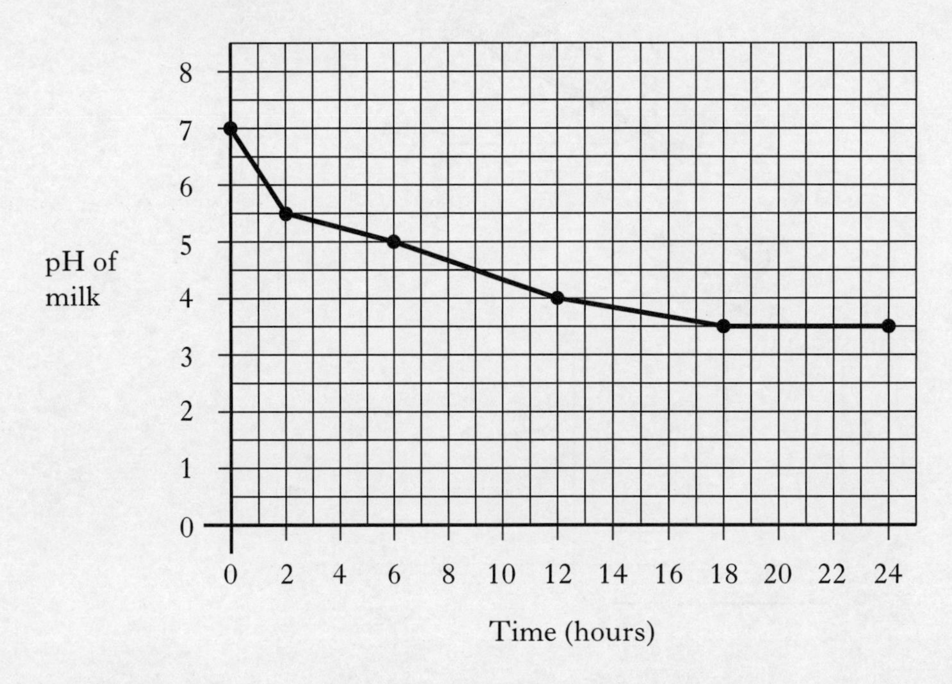

18. The change in pH is caused by the production of acid by

A fungi

B viruses

C bacteria

D yeast.

19. During which period of time was there the greatest change in pH?

A 0–6 hours

B 6–12 hours

C 12–18 hours

D 18–24 hours

20. The table below shows the results of a resazurin test on four milk samples.

Milk sample	*Colour after 30 minutes*
A	pink
B	purple
C	mauve
D	white

Which milk sample contains most bacteria?

21. The solid formed when protein is clotted during cheese making is

A rennet

B curds

C whey

D yoghurt.

Questions 22 and **23** refer to the following graph. This shows the changes in oxygen concentration before and after waste whey is released into a river.

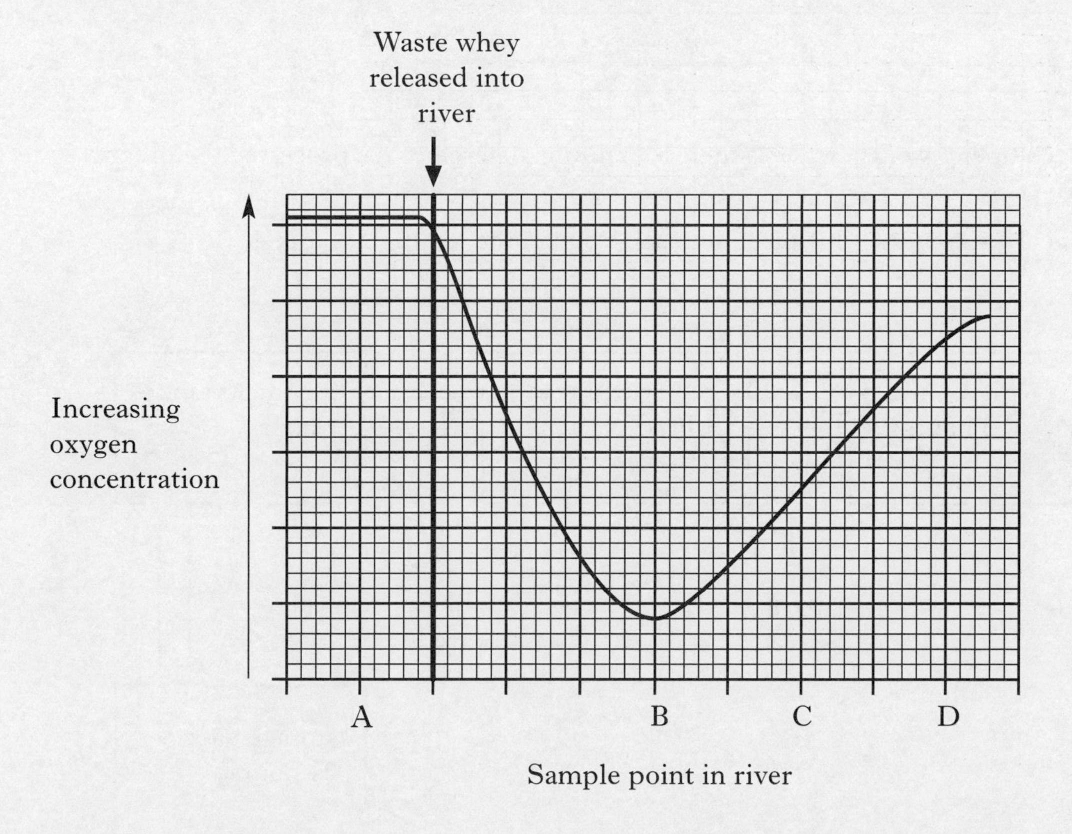

22. At which sample point in the river was the oxygen concentration highest?

23. After waste whey is released into the river, the oxygen concentration

A increases then stays the same

B increases and then decreases

C decreases then stays the same

D decreases and then increases.

[Turn over

24. Which line in the table below shows correctly the effect of antibiotics on the growth of bacteria and viruses?

	Effect of antibiotics	
	Bacteria	*Viruses*
A	✓	✓
B	✗	✓
C	✗	✗
D	✓	✗

Key

✓ = stops growth

✗ = does not stop growth

25. At a sewage treatment works, waste water containing detergents is processed before being released into rivers.

The diagram below shows the time taken for each stage in the treatment process.

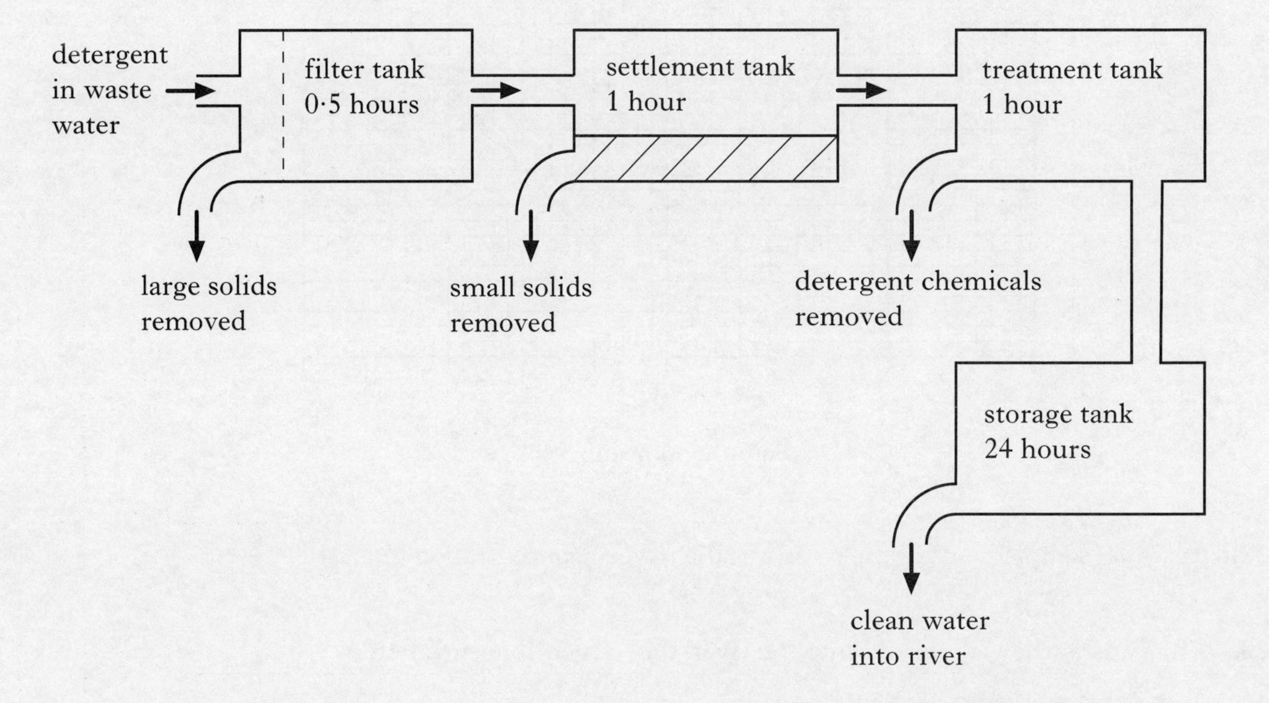

How long does it take to remove all solids and detergent chemicals from the water?

A 26·5 hours

B 21·5 hours

C 2·5 hours

D 1·5 hours

Candidates are reminded that the answer sheet for Section A MUST be returned <u>inside</u> this answer book.

SECTION B

All questions in this Section should be attempted.
All answers must be written clearly and legibly in ink.

Marks | DO NOT WRITE IN THIS MARGIN

1 (*a*) **Read the following passage carefully.**

Second-hand Tobacco Smoke

Second-hand tobacco smoke is the smoke breathed in by non-smokers when other people are smoking. It is sometimes called environmental tobacco smoke and breathing it in is known as passive smoking. Only 1 in 5 of the population smokes but almost everyone breathes in second-hand tobacco smoke at times.

Most non-smokers dislike second-hand tobacco smoke. They complain that it causes headaches, coughs, feelings of dizziness and sickness. It can also cause irritation of the nose, throat and eyes. The smell of tobacco smoke clings to hair, clothes and furnishings.

A burning cigarette is like a mini chemical factory. The smoke contains thousands of chemicals. The smoker breathes in only 15 percent of the smoke from a cigarette. This is called mainstream smoke. The other 85 percent, known as sidestream smoke, goes straight into the air. Sidestream smoke is unfiltered and contains higher concentrations of toxic chemicals than mainstream smoke.

Use information from the passage to answer the questions below.

(i) What name is given to breathing in second-hand smoke?

________________________________ 1

(ii) Name **two** effects that second-hand smoke has on non-smokers.

1 ________________________________

2 ________________________________ 1

(iii) What percentage of smoke from a cigarette is mainstream smoke?

________________________________ 1

(iv) Why is sidestream smoke more dangerous than mainstream smoke?

________________________________ 1

(*b*) Name a harmful chemical found in tobacco smoke.

________________________________ 1

DO NOT WRITE IN THIS MARGIN

Marks

2. Peak flow rate is measured using the instrument shown below.

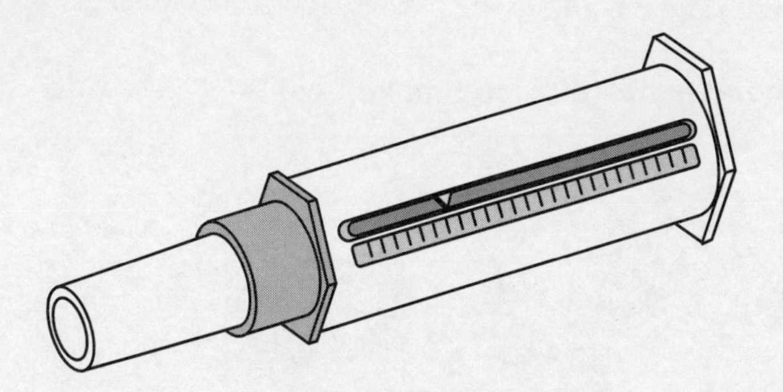

Three peak flow readings of a fourteen-year-old student were taken.

Reading	*Peak Flow* (litres per minute)
1	500
2	510
3	490

(*a*) (i) What is this student's peak flow rate?

______________ litres per minute **1**

(ii) Name **one** factor, other than age, which can affect peak flow rate.

____ **1**

(*b*) **Underline** one option in each set of brackets to make the statement below correct.

Peak flow is the {minimum / maximum} rate at which air can be forced {into / out of} the lungs. **1**

(*c*) Name a medical condition which can be diagnosed and managed using a peak flow meter.

____ **1**

DO NOT WRITE IN THIS MARGIN

Marks

3. A student, when at rest, measured her heart rate three times using a stethoscope and a stopwatch.

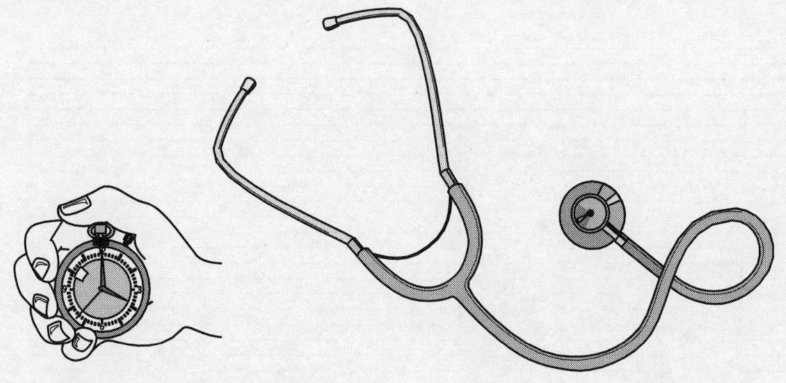

The results are shown in the table.

Measurement	*Number of beats in 20 seconds*
1	21
2	21
3	24

(*a*) Calculate the student's:

(i) average heart rate in 20 seconds;

Space for calculation

______________ beats per 20 seconds **1**

(ii) average pulse rate in beats per minute.

Space for calculation

______________ beats per minute **1**

(*b*) The student then exercised for 30 minutes.

What effect would this have on her pulse rate?

____________________________________ **1**

(*c*) What term is used to describe the time taken for the pulse rate to return to normal after exercise?

____________________________________ **1**

DO NOT WRITE IN THIS MARGIN

Marks

4. (*a*) The bar graph below shows the daily energy requirements of four male students.

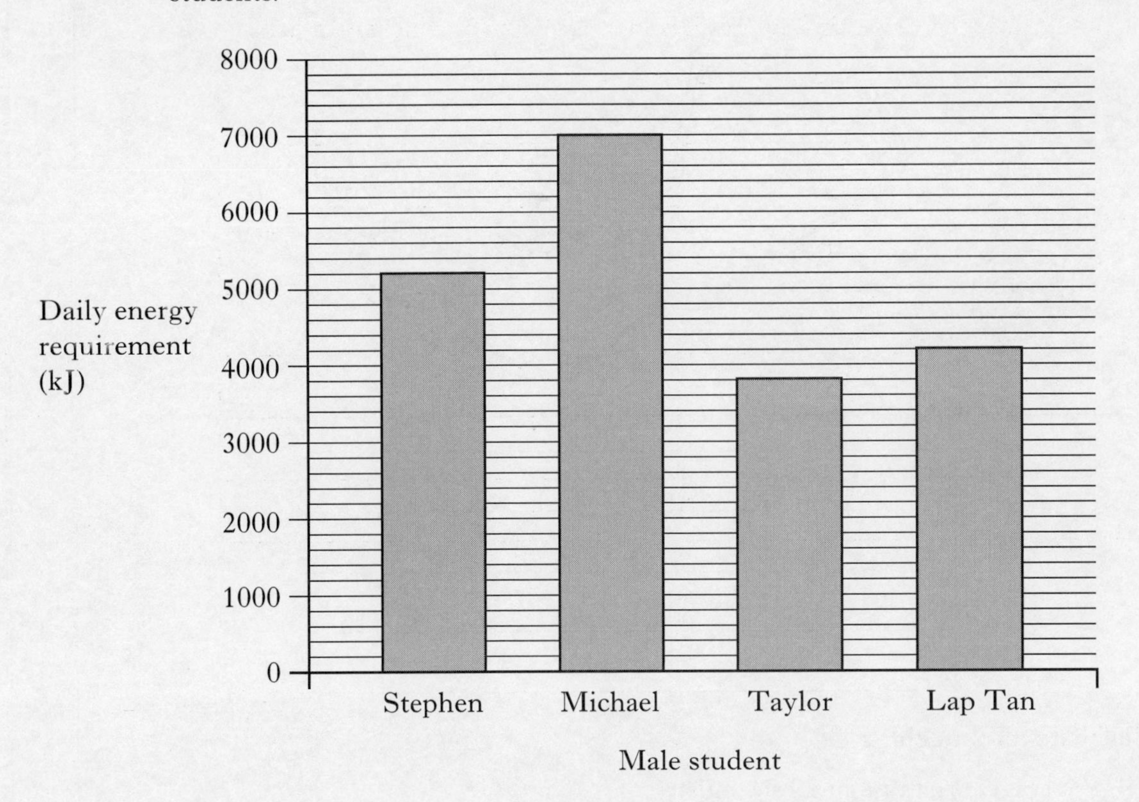

(i) Use the information in the bar graph to complete the table below.

Male student	*Daily energy requirement* (kJ)
Stephen	
	7000
Taylor	
Lap Tan	

2

(ii) Stephen and Michael are the same age and weight.

What evidence from the bar graph shows that Michael is most likely to be an athlete?

____

____ **1**

DO NOT WRITE IN THIS MARGIN

Marks

4. (continued)

(*b*) Tick (✓) **one** box for each food group in the table to show its **main** use.

One food group has been completed for you.

Food group	*Main use*		
	Energy	*Growth and repair of cells/tissues*	*Protection against disease*
Carbohydrates	✓		
Proteins			
Fats			
Vitamins and minerals			

2

[Turn over

DO NOT WRITE IN THIS MARGIN

5. A student carried out an investigation into the effect of watering on spider plants.

Five plants were placed under identical conditions and regularly watered over a period of four months.

Each plant received a different volume of water.

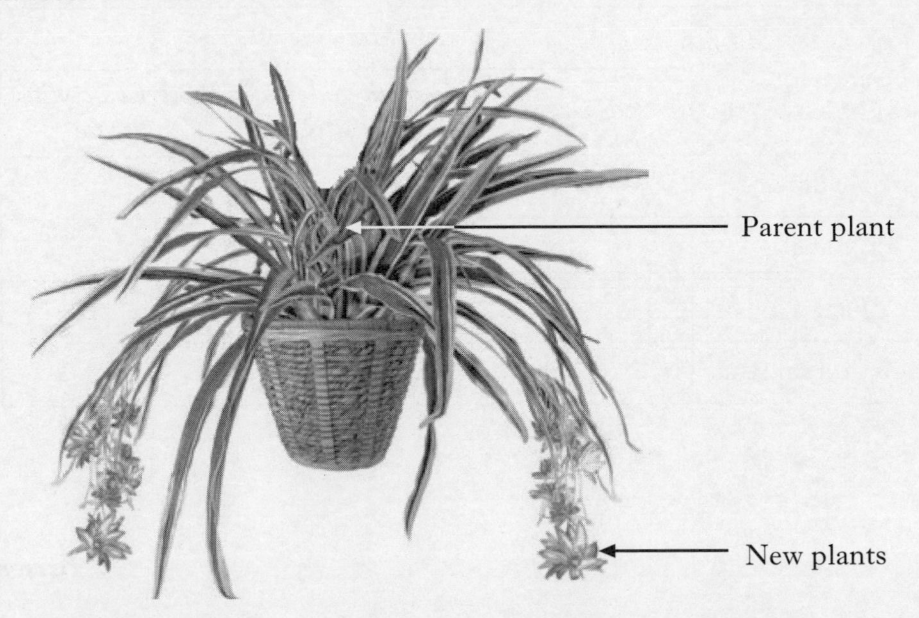

The number of new plants growing from each parent plant is shown below.

Plant	*Volume applied at each watering* (cm^3)	*Number of new plants produced*
A	20	0
B	40	2
C	60	6
D	80	4
E	100	3

DO NOT WRITE IN THIS MARGIN

Marks

5. (continued)

(*a*) On the grid below, complete the **line graph** to show the number of new plants produced by

(1) providing a label for the vertical axis **1**

(2) putting a scale on the vertical axis **1**

(3) plotting the results. **1**

(Additional graph paper, if required, will be found on page 28.)

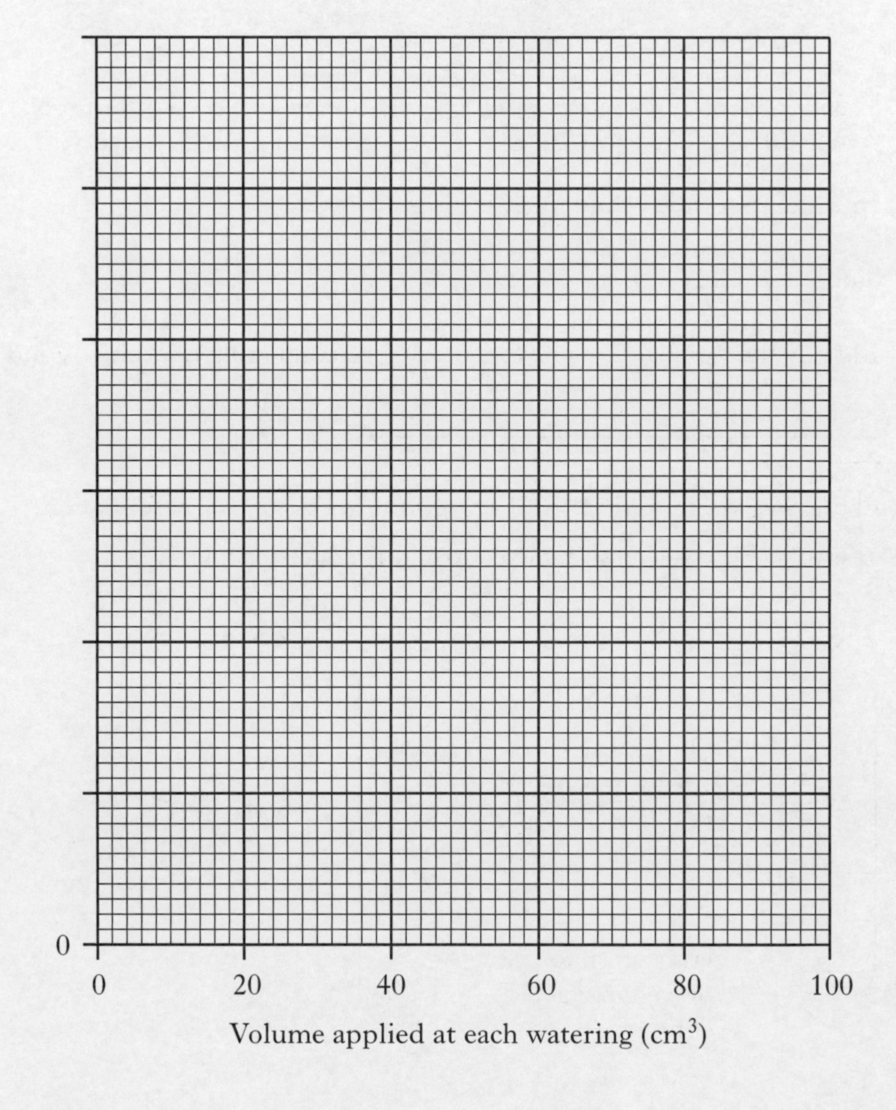

(*b*) Identify the volume of water applied at each watering that resulted in most new plants.

______________ cm^3 **1**

(*c*) Suggest an improvement to the investigation which would make the results more **reliable**.

____ **1**

DO NOT WRITE IN THIS MARGIN

Marks

6. (*a*) The diagram below shows test tubes set up to investigate the conditions required for seed germination.

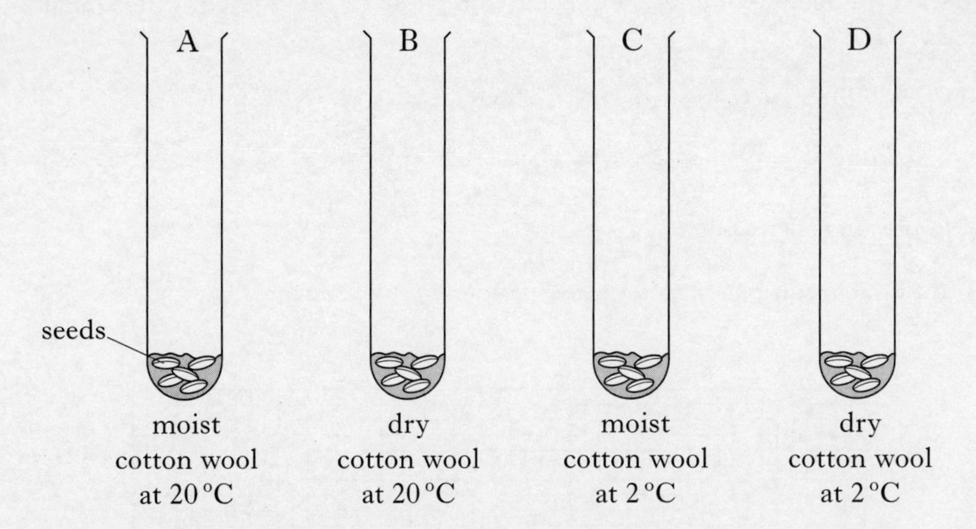

(i) In which test tube would most seeds germinate?

Tube ________ **1**

(ii) Identify the variable being investigated when comparing test tubes A and C.

____________________________________ **1**

(*b*) Test tube X was set up to find out if **oxygen** is necessary for germination.

Label test tube Y to show a suitable control for this experiment.

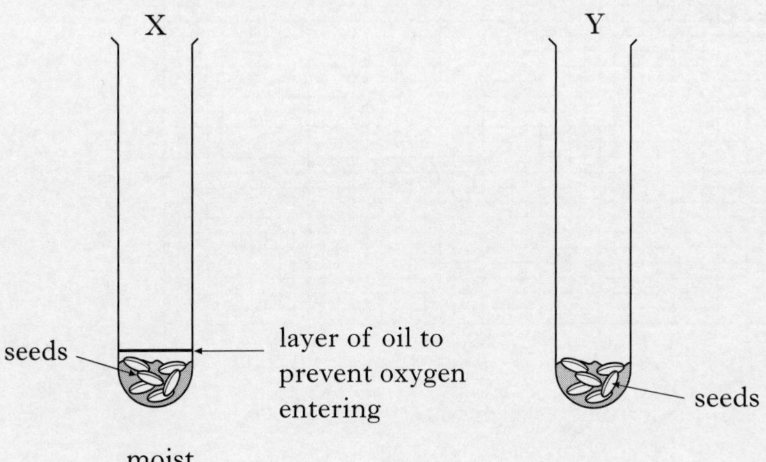

1

DO NOT WRITE IN THIS MARGIN

Marks

6. (continued)

(*c*) Name the part of a seed which is used to provide energy for germination.

____ **1**

(*d*) Some seeds will not germinate until spring when the soil temperature rises.

What name is given to this delay in germination?

____ **1**

[Turn over

DO NOT WRITE IN THIS MARGIN

Marks

7. The table shows the rooting success of cuttings of four varieties of *Fuchsia*.

Variety of Fuchsia	*Successful rooting* (%)
Pink Goon	85
Son of Thumb	65
Brutus	60
White Pixie	70

(*a*) On the grid below, complete the bar graph by

(1) putting a label on the vertical axis **1**

(2) plotting the results for the other varieties. **1**

(Additional graph paper, if required, will be found on page 28.)

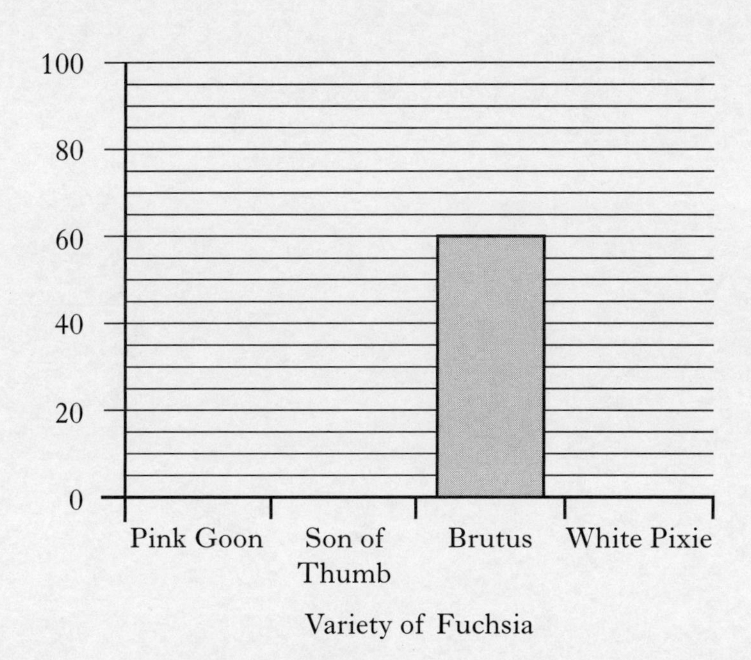

(*b*) Which variety of *Fuchsia* was least successful at rooting?

____ **1**

(*c*) What could improve the successful rooting of the cuttings?

____

____ **1**

DO NOT WRITE IN THIS MARGIN

Marks

8. Commercially grown plants, such as lettuce, are often grown in a polythene tunnel.

(*a*) State **one** reason why plants are cultivated in this way.

______________________________ 1

(*b*) Disease and pests must be controlled when growing plants.

State **one** way in which aphids and grey mould can be controlled.

Aphids ______________________________ 1

Grey mould ______________________________ 1

(*c*) Name the process by which plants produce food for growth.

______________________________ 1

[Turn over

DO NOT WRITE IN THIS MARGIN

Marks

9. (*a*) The table below shows the components of two soft cheeses.

	Component (g per 100 g)			
Type of cheese	*Protein*	*Carbohydrate* (including sugars)	*Fat*	*Water*
Cottage cheese	15	2	4	79
Cream cheese	3	0	48	49

(i) Calculate the ratio of fat in cottage cheese to that in cream cheese.

Express your answer as a **simple whole number ratio**.

Space for calculation

_____________ : _____________ **1**

Cottage cheese   Cream cheese

(ii) Use the information for cottage cheese in the table to label the pie chart.

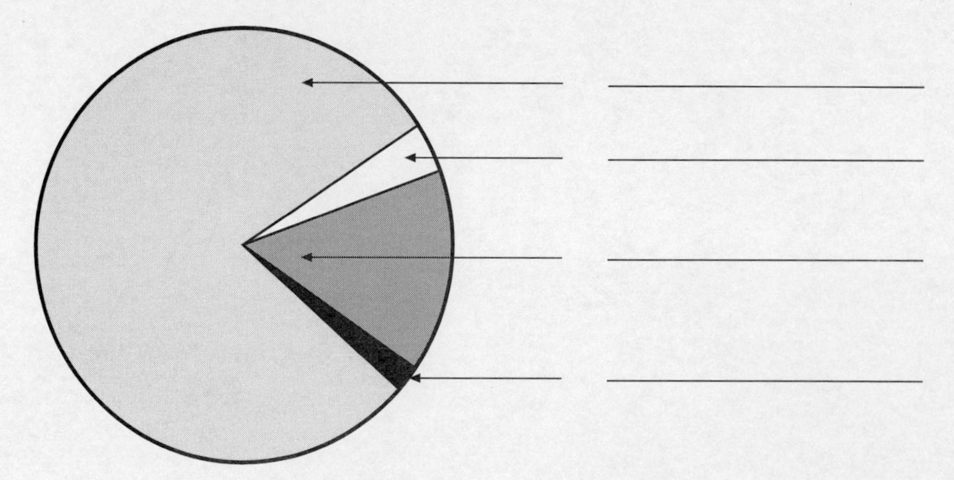

1

DO NOT WRITE IN THIS MARGIN

Marks

9. **(continued)**

(*b*) A substance is added to milk during cheese making which makes the protein clot.

(i) Name this substance.

______________________________ 1

(ii) One source of this substance is genetically engineered yeast.

State **one** other source of the substance.

______________________________ 1

(*c*) Whey is a waste product of cheese making.

(i) Name the organism involved in the upgrading of waste whey.

______________________________ 1

(ii) Give **one** example of a product made from whey.

____ 1

[Turn over

DO NOT WRITE IN THIS MARGIN

Marks

10. (*a*) A student carried out an investigation to compare biological and non-biological detergents.

Four pieces of cloth were stained and then each washed using a different detergent.

The time taken for each stain to disappear is shown in the table below.

Detergent	*Type of detergent*	*Time taken for stain to disappear* (minutes)
Alpha	Biological	100
Beta	Non-biological	140
Gamma	Biological	120
Delta	Non-biological	160

(i) What conclusion can be drawn from these results?

____

____ **1**

(ii) Which variable was altered in this investigation?

____ **1**

(iii) Identify **two** variables which should have been kept the same when setting up the investigation.

1 ____

2 ____ **2**

(*b*) Explain why using biological detergents is claimed to save energy.

____

____ **1**

DO NOT WRITE IN THIS MARGIN

Marks

10. (continued)

(*c*) Biological detergents contain enzymes.

(i) Which type of living organism is used to produce these enzymes?

______________________________ 1

(ii) Some people are allergic to the enzymes in biological detergents.

Name **one** medical condition which can be caused by this allergic reaction.

______________________________ 1

(iii) How does the manufacturer reduce the chance of the enzymes in detergents causing an allergic reaction?

______________________________ 1

[*END OF QUESTION PAPER*]

SPACE FOR ANSWERS

DO NOT WRITE IN THIS MARGIN

ADDITIONAL GRAPH PAPER FOR QUESTION 5(*a*)

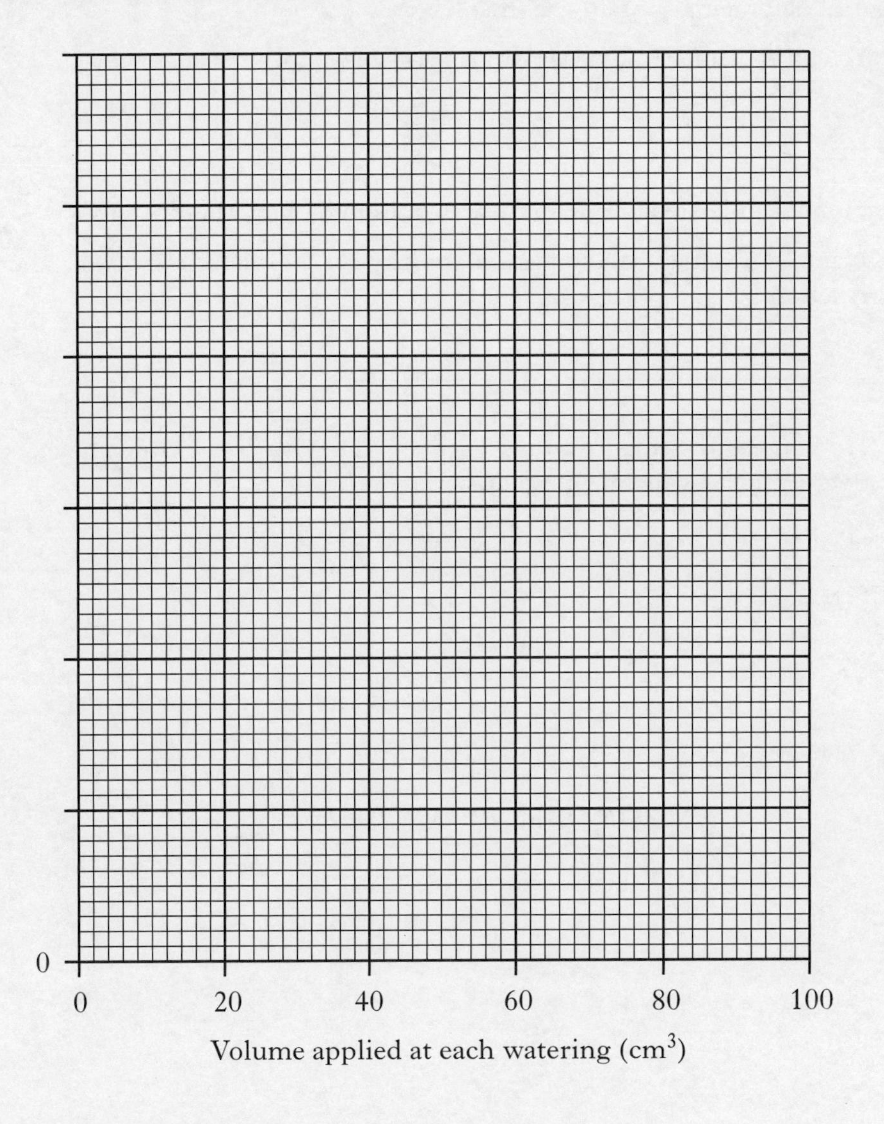

ADDITIONAL GRAPH PAPER FOR QUESTION 7(*a*)

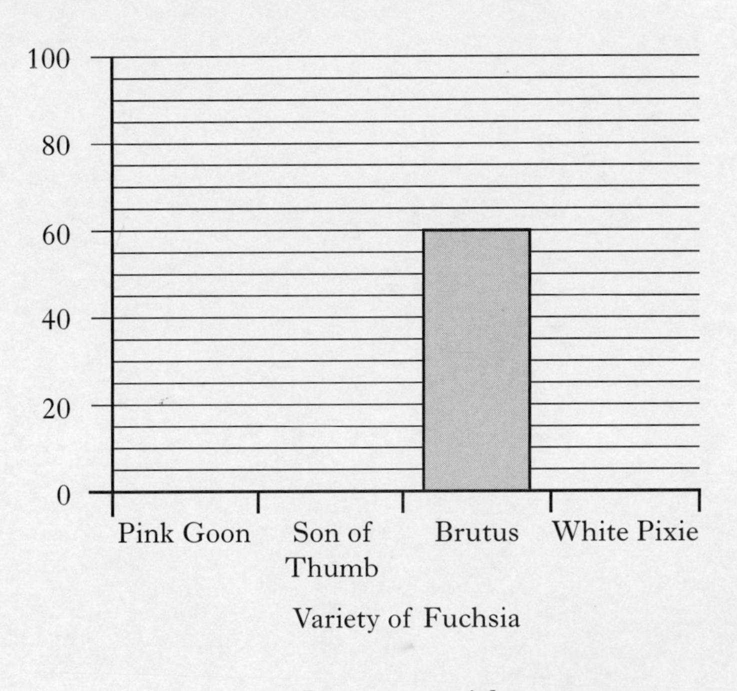

DO NOT WRITE IN THIS MARGIN

SPACE FOR ANSWERS

DO NOT WRITE IN THIS MARGIN

SPACE FOR ANSWERS

SPACE FOR ANSWERS

DO NOT WRITE IN THIS MARGIN

SPACE FOR ANSWERS

[BLANK PAGE]

[BLANK PAGE]

[BLANK PAGE]

[BLANK PAGE]